Howard E.

Howard Florey

Something
· Spectacular

Something Spectacular

MY GREAT LAKES SALMON STORY

Howard A. Tanner

MICHIGAN STATE UNIVERSITY PRESS | EAST LANSING

☉ The paper used in this publication meets the minimum requirements of
ANSI/NISO Z39.48-1992 (R 1997) (Permanence of Paper).

MICHIGAN STATE UNIVERSITY PRESS
East Lansing, Michigan 48823-5245

Printed and bound in the United States of America.

28 27 26 25 24 23 22 21 20 19   1 2 3 4 5 6 7 8 9 10

LIBRARY OF CONGRESS CATALOGING-IN-PUBLICATION DATA
Names: Tanner, Howard A., author.
Title: Something spectacular : my Great Lakes salmon story / Howard A. Tanner.
Description: East Lansing : Michigan State University Press, [2018]
| Includes bibliographical references and index.
Identifiers: LCCN 2017061031| ISBN 9781611863031 (cloth : alk. paper)
| ISBN 9781609175825 (pdf) | ISBN 9781628953473 (epub) | ISBN 9781628963472 (kindle)
Subjects: LCSH: Coho salmon fishing—Great Lakes (North America)
| Coho salmon fisheries—Great Lakes (North America)
Classification: LCC SH686.4 .T36 2018 | DDC 639.2/7560977—dc23
LC record available at https://lccn.loc.gov/2017061031

Book design by Charlie Sharp, Sharp Designs, East Lansing, MI
Cover design by Shaun Allshouse, www.shaunallshouse.com
Cover image of Howard Tanner is used with permission
from Martin Rogers/National Geographic Creative.

Michigan State University Press is a member of the Green Press Initiative and is committed to developing
and encouraging ecologically responsible publishing practices. For more information about the Green
Press Initiative and the use of recycled paper in book publishing, please visit www.greenpressinitiative.org.

Visit Michigan State University Press at *www.msupress.org*

To Helen,
my loving, caring, sharing,
and supportive wife of 70+ years.

To Dr. Robert C. Ball,
my educational/professional
mentor and friend.

CONTENTS

ix   FOREWORD, *by John L. Hesse*

xiii   ACKNOWLEDGMENTS

xvii   INTRODUCTION

DEVELOPING A FIRM FOUNDATION

3   Growing Up with the Fishes

11   The "Greatest Generation" Goes to War

24   Building the Educational Framework

38   Professional Practitioner

THE GREAT LAKES CONTEXT

57   The Inland Seas

64   Human History and the Great Lakes Fishery

72   Fisheries Management

MICHIGAN MATTERS

79   Meanwhile in Michigan

85   Returning to Michigan

91   Commercial Fishing

98   The Great Lakes Fishery Commission

101   Tribal Fishing Rights

FISHES OF THE GREAT LAKES

107 Lake Trout
113 Sea Lamprey
120 Alewife
124 Salmon

INTRODUCING COHO

137 The Call
140 A New Day Dawns
145 Supporting and Opposing Forces
161 Eggs from Oregon
165 The Hatchery Situation

PREPARING TO LAUNCH

173 Financial Challenges
176 Making the Case
182 New Personnel

THE FIRST RELEASE

187 April 2, 1966—the Official Beginning
190 The Summer of 1966
194 The Jack Run

PROMISE FULFILLED

201 Alewives Reprise
205 The Dream Come True and a Nightmare
217 Swimming Upstream
221 Sustaining the Excitement

FURTHER DEVELOPMENTS

229 Stewardship and Contaminants
232 Economics

239 CONCLUSION
246 NOTES
260 INDEX

FOREWORD

The story is spectacular; the storyteller is extraordinary. *Something Spectacular: My Great Lakes Salmon Story* gives the reader insights into Michigan's salmon legacy that are inspiring, knowledge-based, and entertaining. I've known Dr. Howard Tanner for nearly fifty years, first meeting him while I was a student at Michigan State University and continuing as an employee when he was director of the state's Department of Natural Resources (DNR). More recently, in retirement, we have been fishing partners and close personal friends. From reading this book, you will come to share my admiration for Howard as a catalyst in the enhancement and preservation of the Great Lakes as the incredible resource they are today.

Dr. Tanner is often referred to as the "father of the Great Lakes salmon fishery." As he was inducted into the American Fisheries Society Hall of Excellence in 2008, his early contributions in shaping the Great Lakes salmon stocking program were referred to as "the most significant biomanipulation program in the history of fisheries management." The detailed accounts in this book describe how this all came about.

Even in his nineties, Howard remembers the situations in his youth that pointed him toward the fine fisheries professional he became. Over the years, as we've fished the Great Lakes and inland waters of this beautiful state, I've heard many of the stories he has told in this book. It's wonderful to finally see them in greater detail and in print. Howard is an amazing storyteller, and readers of the book will delight in his incredible life story, his humor, his professionalism, his sense of values, and his contributions, not only to Michigan but to the entire Great Lakes Basin.

Fisheries professionals, anglers, students, and others who read this book will benefit from the knowledge they will gain about the Great Lakes ecosystem and how it was developed into a world-class fishery valued at more than $7 billion per year to our economy.

As a student working toward graduate degrees in fisheries at Michigan State University, I was privileged to serve with the first crew of the S/V *Steelhead* in 1967–68, working out of Charlevoix. In the book, you'll learn about how this vessel came to exist (somewhat illegally) under Howard's direction. It played an important early and ongoing role in evaluating the status of fish stocks in Lake Michigan, including gathering critical data about movement patterns and life history of the newly introduced salmon.

Coincident with the 2016 fiftieth anniversary of introducing salmon into the Great Lakes, Howard's contributions to this fishery were recognized by naming a new DNR research vessel the R/V *Tanner*. With state-of-the-art research tools, the *Tanner* is assigned to evaluate Lake Huron fisheries as the *Steelhead* has in Lake Michigan for half a century. Appropriately, Howard and his lovely wife Helen participated in the new ship's christening and dedication. In this book, you'll sense how Helen's constant support and love for more than seventy-three years throughout his education and career contributed to his success.

What you will not read in this book are some of Dr. Tanner's other accomplishments while director of the Department of Natural Resources from 1975 through 1983—a critical time in Michigan's rise as a conservation leader. By 1975, Michigan had evaluated and dealt with a major mercury crisis and discovered another major contaminant in Great Lakes fish—polychlorinated biphenyls (PCBs)—and we had identified many large point sources of this toxin.

Dr. Tanner quickly recognized the futility of controlling PCBs one discharge at a time. He invited me to attend a governor's cabinet meeting to recommend that Michigan take the lead to ban further use of this dangerous industrial chemical. It had contaminated our Great Lakes fish and was contributing to reproductive failure of our eagles and other fish-eating birds, as well as threatening human health. Governor Milliken and his cabinet responded positively to the arguments presented by Howard, and Michigan soon led the nation in banning PCBs.

Over the ensuing years, Howard and I testified at congressional hearings and national meetings about PCBs and other contaminants, and I learned a great deal about his skills in influencing national and state policy on environmental issues. Those were the same skills that had earlier brought his success at turning the Great Lakes into a world-class sport fishery through introducing salmon.

Even in retirement, Howard continues to wield influence on major environmental issues that he sees threatening the Great Lakes. He consults frequently with state leadership and participates with other state retirees in an advisory group called Michigan Resource Stewards. In 2016, he testified vigorously at state hearings against the permitting of net pen aquaculture in Great Lakes waters, which he and other environmentalists see as a huge threat to resident fish populations and water quality. He has also been very vocal about the need for a permanent solution to prevent the entry of the invasive Asian carp species into Great Lakes waters.

Through this book, it will become obvious that Dr. Tanner loves Michigan, and that his boldness and perfect timing played a critical role in making its Great Lakes fishery truly "Something Spectacular."

ACKNOWLEDGMENTS

First and foremost, I thank my wife Helen for her loving companionship and support. She also had the education in zoology to share in many aspects of my professional activities. While I was in the Army overseas during World War II, she guided and developed my educational choices to focus on freshwater fisheries and to attend Michigan State University. During my MSU graduate fieldwork, she helped collect specimens, kept all the records in duplicate, and tutored me in foreign languages by the campfire. She sustained our growing family of three sons and our household—from student days at MSU, through twelve years in Colorado, to settling back in Michigan. She has shared in the preparation of this book through its several iterations and encouraged me to see it through.

I also thank Dr. Robert Ball, who, through his undergraduate classes, opened for me a full understanding of and interest in freshwater environments, lakes, streams, and the Great Lakes. His guidance through both of my graduate programs gave me the knowledge, expertise, and skill to manage fisheries in Colorado and Michigan. His tip about the availability of coho salmon eggs was a crucial turning point in the development of the Great Lakes sport fishery.

It was one thing to initiate the effort to introduce Pacific salmon to the Great Lakes. It was quite another to take the reins and manage Michigan's fisheries for a significant, sustainable future. I especially appreciate the role that Dr. Wayne Tody played so well for many years, first in partnering in and implementing the plan for stocking cohos, and subsequently in developing and leading the chinook fishery.

I also want to thank publicly those Michigan Conservation Department leaders—especially director Ralph MacMullan and deputy director Chuck Harris—who challenged us to do this "spectacular" thing and supported us at critical times.

For decades, people told me that I must write the story of how the Great Lakes salmon fishery came to be. In the late 1990s, Wayne Tody and I collaborated on a chapter for a book published by the American Fisheries Society, but that account for a largely professional audience didn't tell the whole story.

Finally, in the early years of the twenty-first century, while leading a project about Michigan's Chronic Wasting Disease (in deer and elk), I was fortunate to work with Mrs. Betsy Clark. After hearing some of my accounts of those early events, Betsy urged me to write the book and promised her help. I had never written anything more extensive than a magazine article or book chapter, but her encouragement and persistence proved to be one of several turning points in this undertaking. Not only did she transcribe my handwriting electronically, she and her husband Gary provided extraordinary practical and personal support at a time when I was losing my eyesight and finding it increasingly difficult to read and write. She also located Dr. Michael VanDyke, associate professor of English at Cornerstone University, whom I appreciate for editing the manuscript quite professionally. I can't thank Betsy enough for all she contributed to helping develop this book.

Not finding a receptive publisher for the initial iteration, I worked with editor Jim Bedford to develop a different approach. I really appreciate Jim's efforts, and have been able to use some of that material in this iteration.

As my eyesight has deteriorated, it has been most helpful to have technological devices and software to help me with dictation and "reading" the drafts and e-mail messages about the book. Connie Wolf has been a ready tech support person, and I very much appreciate her assistance with these matters.

Finally, Jim Bedford and another former DNR colleague, William Murphy—both published authors—led me to contact Carol Swinehart, an award-winning writer, editor, and producer of materials about the Great Lakes and the fishery in particular. Her guidance, research, and editing has helped produce a format that combines documented facts and figures with my specific memories of events and developments in the story of how Michigan led the creation of the Great Lakes sport fishery by introducing Pacific salmon. To her I express my most sincere and deepest thanks.

I also appreciate MSU Press's commitment to publishing this work. Acquiring editor Julie Loehr, editor Annette Tanner, and the editorial, production, and marketing teams have greatly assisted this novice author. Thanks also to those who assisted with acquiring suitable images, especially Gary Whelan, Elyse Walter, and Dave Kenyon (Michigan DNR); Linda Garcia (University of Michigan Museum of Zoology); Peter Thompson (biological illustrator); Ted Lawrence (Great Lakes Fishery Commission); and Gina Martin (National Geographic Creative).

INTRODUCTION

The introduction of Pacific salmon into the Great Lakes, and the world-class sport fishery that has since evolved have been described in glowing terms. The scientific community has categorized the achievement as "the largest and most successful bio manipulation ever attempted."[1] The *Saturday Evening Post* titled its Fall 1972 story "The Miracle of the Fishes."[2] In an article titled "Michigan: The State That Almost Wasn't," a *Reader's Digest* author proclaimed that the "restoration of life to the Great Lakes . . . is the most significant accomplishment for the state."[3]

The Great Lakes sport fishery now extends from the north shore of Lake Superior to the outlet of Lake Ontario. Economists have reported that nearly 1.7 million people are angling on the lakes, and that this effort is producing as much as $7 billion in economic activity annually in the United States.[4]

Fifty years ago, this result was hardly imaginable. At that time, in the fall of 1964, I was the new chief of the Michigan Conservation Department's Fish Division. Challenged by the department's director to do something "spectacular," it was my vision, my decision, my accomplishment to initiate, lead our staff to develop, and begin to implement a plan for enhancing the state's recreational fishery. In this book, I share how I led this initiative through its early stages by obtaining sufficient coho salmon eggs, arranging for rearing the young salmon to their release date, dealing with "naysayers," and overcoming budget and other political obstacles, including the outright opposition of the federal government.

Several factors were of decisive or crucial importance in the success of this magnificent undertaking. Some were important precursors; others were instrumental as we proceeded; still others were supportive. I will highlight the **critical factors** listed below at appropriate places, and I hope that featuring them will help you understand how we succeeded in establishing the world's greatest freshwater sport fishery.

- Introducing coho salmon in Colorado
- The Great Lakes as suitable habitat
- The development of fisheries management as a science
- Increasing public demand for greater fishing opportunities
- Sweeping changes in Michigan government
- The work of the "blue ribbon" committee
- Changing the key value in Michigan's Great Lakes fishery
- The modest success of pink salmon in Lake Superior
- The "accidental" alewife's population explosions in the Great Lakes
- Gaining control over sea lamprey in the upper Great Lakes
- Ohio's refusal to yield its management authority to the Great Lakes Fishery Commission
- Relationships with West Coast fisheries biologists
- Oregon's Moist Pellet food
- New Michigan Conservation Department leadership
- Public support for the salmon plan.

Since those early years, many other people, usually led by Michigan's Fisheries Division, have nurtured and expanded this fishery throughout the Great Lakes Basin. Chief among them was my immediate successor, Dr. Wayne Tody, who enlarged the plan to include chinook salmon and implemented it with great professionalism, skill, dedication, and success over many years.

However, the full story of how we created this extremely valuable fishery has never been told. The background that we brought, the obstacles that we surmounted, the boldness of our decisions—based on our absolute understanding and certainty of the suitability of this introduction into the world's largest freshwater environment—need to be described. Very few people ever possessed the full knowledge of how we achieved this monumental goal, and few of those are living today.

The sum of my previous life experiences had prepared me to make the right decision at the right time and then to overcome all the difficulties that we would encounter. Therefore, I'm very pleased that I now have this opportunity to tell what I know and experienced. I hope that, having read my account, you will have a greater understanding and appreciation of the Great Lakes salmon fishery.

Developing
a Firm Foundation

Growing Up with the Fishes

My ties to fishing began at a very early age. My father and mother moved to Mancelona, Michigan, in 1925, when I was just two years old. The circumstances and opportunities of life in that part of Michigan in that small town at that time were basic to my early development.

My father loved to fish and provided me with wonderful fishing experiences. He fished for one principal species—the brook trout. The brook trout is a particularly beautiful fish. When living, its colors can be truly brilliant, and even hours after death, when it's washed and prepared for meals, it's still beautiful.

I remember this quite vividly, though I think that I was no more than three years old in the beginning. It was my privilege to put all my fondest brook trout (in those days the limit was twenty-five) into a dishpan. Then I would lay them out on a piece of newspaper and arrange them according to size. I liked the looks of them and I liked the feel of them, so soft and smooth. Those early memories account for the beginnings of my love for fishing. I loved the brook trout and then the habitat, the environment, the natural surroundings that were part of a brook trout fishing expedition. The season ran from May 1 until Labor Day. We also fished perch in the wintertime, but in my very early years, perch felt rough and had spines and were far less desirable than brook trout.

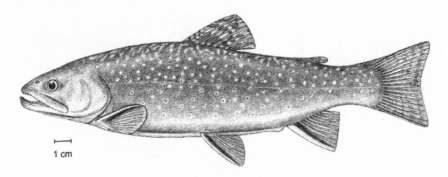

1 cm

Brook Trout—my favorite fish as a child.
AN ATLAS OF MICHIGAN FISHES, USED WITH PERMISSION FROM THE UNIVERSITY OF MICHIGAN MUSEUM OF ZOOLOGY

Brook trout fishing at that time in that place was a challenging and lovely experience. The streams were small, cold, and clear. In my early fishing days, I didn't have boots or waders. I would walk the banks of selected portions of the Cedar River between Mancelona and Bellaire. It was in the upper Jordan River valley in the area that we knew as the "green swamp," so named because it was occupied by a sizable stand of cedar and tamarack. This peninsula was shaped almost like a hairpin, with the river nearly encompassing it, and had not been burned by the frequent forest fires.

Brook trout are nearly always to be found in deep water under submerged logs or under bridges, logs, or beneath undercut banks. Only in certain areas are they found out in the open.

We fished with a metal rod, black cotton line, and a double-bladed spinner loaded with a sculpin, which we referred to as a "muddler." It took considerable skill to cast the spinner accurately so that the line would float gently down underneath the log or the bank. In those days, brook trout had to be seven inches long to keep, and we would often catch several undersized brook trout before we got a keeper. By the time that I was old enough to think about exceeding the limit—twelve years or so—the limit had been reduced from twenty-five to fifteen per day.

Brook trout are really delicious—cooked in a frying pan, usually with lard or bacon grease, with its head on. We ate the muscle in the cheeks, and I still love the crispy tail fin. Brook trout were the center point for a matrix of my experiences in the out-of-doors. My father loved everything to do with the outdoors, and he was my role model in that and in so many other ways.

One summer, Dr. Jan Metzlaar from the Michigan Conservation Department visited us.[1] He was surveying Michigan's inland lakes, examining and identifying their fish populations. He was given my father's name as somebody who fished a great deal. He came and stayed with us for two or three days, parking one of the early departmental panel trucks in our yard. The truck had racks along the insides with glass jars full of preserved fish of various species.

Probably the significant element of his visit was a remark that my father made in my presence several times, words to the effect, "Just imagine! This man is well paid for going and examining the fish populations of Michigan lakes!" I think that my father's remarks revealed that he would have loved to have had the opportunity to do what Metzlaar was doing. Somewhere in my mind, I thought, "My father can't do that, but perhaps I could"—sometime in the distant future.

The Jordan River

In the summer of 1937 my father opened a meat market. We had a wonderful summer of fishing together, usually on the Jordan River on Sunday morning or late Sunday afternoon—his only day off. A railroad had been built on the floor of the valley, the timber removed and the valley repeatedly burned over. The roads we used were frequently built atop the old railroad grade with some of the ties still in place. One entrance took us through a farmer's gate, which we carefully closed because his cattle still roamed and grazed in places where the second growth of trees had not yet begun. We went there to fish brook trout, but we also went to hunt mushrooms in the spring and grouse (we called them partridge) in the fall. In the spring, we would also make a trip to pick trailing arbutus.

I relished the sights and sounds of the valley in the early morning. On various occasions, as I moved quietly along this stream, I saw a mink—that's right, M-I-N-K. We occasionally fished the impounded tributary dammed by the beaver, and occasionally I would see one. I sometimes saw a great blue heron stalking a green frog, and admired the unbelievable beauty of the first male wood duck that I saw. These were all part of those early trips with my father to the Jordan River valley.

It was on one of those trips fishing for brook trout that I caught a rainbow trout just over twenty-five inches long. What a surprise! I still vividly recall the excitement of catching that fish. The stream was full of logs and other obstacles, and I successfully subdued it to a point where I could heave it toward my father's outstretched

Rainbow Trout—my largest catch as a child.

HOWARD A. TANNER COLLECTION

hands while he stood in the middle of the river. He grasped the fish—he referred to it as a sockdolager (very exceptional)—and tossed it to a grassy spot near where I was standing; then I pounced on it. It was more than thirty years before I caught a larger trout. I think that specific event was important; because of that, I came to consider myself an expert fisherman.

Bellaire Days

Another very important event came in early September of 1937 when we moved to Bellaire because my father had become sheriff of Antrim County. Part of the arrangement was our dwelling next to the jail.[2] If I had enjoyed fishing while living in Mancelona, Bellaire was paradise. A small village, in those days about 450 people, it's located on the Intermediate River—a part of the Antrim County chain of lakes, where almost a dozen lakes are connected by streams and eventually flow into

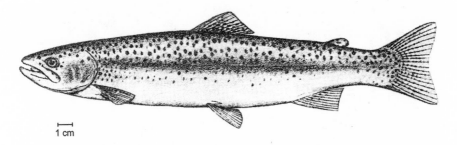

Rainbow Trout.

AN ATLAS OF MICHIGAN FISHES, USED WITH PERMISSION FROM THE UNIVERSITY OF MICHIGAN MUSEUM OF ZOOLOGY

Lake Michigan at Elk Rapids. These lakes, these rivers were heavenly for me. The Intermediate River was almost at our back door. I had a car to drive, and fishing opportunities were abundant all around me.

When I was fifteen, I found a job at Jim's Bait and Tackle Shop. Jim's shop was another facet of my move towards a career in fisheries. I waited on customers; I pumped gasoline; I cleaned and bailed out boats; I caught green frogs and muddlers. Once or twice, I dug night crawlers, which I didn't like.

I also had a printed business card, identifying myself as a fishing guide. Perhaps it was that big rainbow trout that convinced me that I had the expertise to guide fishing trips. From then until the summer of 1942, I worked as a guide. It was a great experience with multiple fishing opportunities—think about it: I got paid for going fishing! At various times during my life, I have described that as the best job I ever had.

I guided men and, once or twice, a woman with her children. I guided people who I considered wealthy. They had great equipment, and often they had a summer home at one of the nearby lakes. In general, they were successful, well-educated,

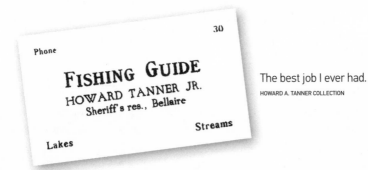

The best job I ever had.

HOWARD A. TANNER COLLECTION

Bud Snook—one of my
fishing guide clients.
HOWARD A. TANNER COLLECTION

well-spoken, very intelligent people from a wide variety of professional experiences. One outstanding individual by the name of Wendell Llewelyn was a vice president of General Motors, heading up the Chevrolet division. He and his family had a summer cottage on nearby Lake Bellaire. We usually fished alone in the late evening, and we had numerous casual conversations of all kinds. He was well-educated and articulate and moved in circles of power with which I had no personal experience. This connection opened avenues of thought for me that expanded my horizons in many ways.

My high school years in and around the small town of Bellaire included further contacts and observations related to one fish or another. I found the annual spawning runs of other fish species fascinating. Less than fifty yards from my back door, in the Intermediate River, were runs of hundreds of big walleyes. Later, at the end of their spawning migration, fishes such as suckers and red horse would pile up by the thousands below the dam immediately upstream, and I would spear some of them. Then our family, the prisoners in the county jail next door, and friends down the street would eat some.

Some of my experiences in the lakes and streams come under the heading of what my father called "penoverating." Roughly translated, that means just looking around to see what is happening. For the first time, I saw members of the sunfish family spawning, the nests that were built, and the male guarding those nests. I found the stickleback's nest-building fascinating—a nest of woven vegetation that encapsulated the eggs and the male of the species as he guarded them.

The peering that we did at night with an underwater light revealed much more than the whitefish or the gar pike that we were spearing. I saw the immense numbers of game fish present in some locations. I saw the huge muskellunge in pairs as they entered the marshes to spawn. My father and I once made a trip to Elk Rapids, where I saw members of the local sportsmen's club seining large numbers of big rainbow trout by torchlight, placing them in washtubs and transporting them above the dam for release into the Elk River. I saw people dipping smelt at night at the mouth of the Jordan and Boyne Rivers. My father was often present, and his clear expressions of interest and excitement guided me to appreciate what he appreciated.

Speaking of a Career

I was president of our high school's junior and senior classes, and I gave what was termed the "president's speech" at our graduation ceremonies. The important point here was that I said, in closing, that I intended to pursue a career as an ichthyologist. That term technically means a person who studies the classification of fishes—I suppose that was close enough for the time. Perhaps I thought some about my future career, but it was a long way from any definition.

During that summer of 1941, having completed high school, my Uncle Hugh (only eight years older than I, and the closest thing I ever had to a brother) was working as an official for the YMCA and stationed in the town of Hastings, Michigan. He and his wife Virginia had graduated from Western Michigan College, and they took an interest in how I might move towards a college education. I visited them in Hastings, and they took me to see Western in Kalamazoo, then to a school in East Lansing known as Michigan State College.

During the visit to Michigan State, I stopped in at the laboratories of Dr. Peter Tack, who taught the fisheries courses. What I saw was another display of preserved fish in glass bottles. Somehow, this wasn't what I had in mind. In 1943, when the Army sent me to Michigan State College to study engineering, I spent almost nine months on campus and never bothered to go over to see Dr. Tack or his fisheries laboratory. I had no firm goal of a career in fisheries then.

The "Greatest Generation" Goes to War

I lived all my young life in the cluster of a middle-class family—my parents, my maternal grandparents, an aunt, and two uncles. I was the only child in a group that included teachers, former teachers, or would-be teachers. My father brought another dimension—love of the out-of-doors, fishing, and hunting—to this family matrix. I lived in two small towns—Mancelona and Bellaire—whose total populations were not quite two thousand people.[1]

In June 1941, the nineteen members of my high school senior class and our chaperones took an eleven-day bus tour of Niagara Falls, New York City, and the historical landmarks of our country in Philadelphia and Washington, DC. It had been a great four years during which I had enjoyed varsity sports and served as class president my junior and senior years. We would never be together like this again. All eight able-bodied boys would go on to serve in the military, and one would die on Okinawa. I spent the summer guiding fishing clients and doing odd jobs on surrounding fruit farms to save money for college. As Tom Brokaw taught me many years later, I had just lived through the decade or more of the Great Depression, and I was about to begin the second phase of experiences that would complete my eligibility for being a member of "the greatest generation."[2]

In the fall, I turned eighteen, and it was from this foundation that I started my college career at Western. I lacked any specific goal at that time, so I enrolled in what was known as a general degree. My mother's encouraging words were, "Please finish at least one year of college." That advice was wise and prophetic.

A Life-Changing Day

The first term was routine, I guess—getting acquainted with college life, living in a dormitory, and occasionally hitchhiking the two hundred miles back home for a weekend. At the end of one such weekend in early December, I was heading back to campus. It was cold, and I was waiting in the swirling snowflakes for my next ride. When I got into the car, the radio was blaring the news that the Japanese had bombed Pearl Harbor.

That news was surprising. However, the ongoing and discouraging reports of the war in Europe—the fall of France, the German armies sweeping across Russia and North Africa—had produced a sustained dread of what lay ahead throughout my high school years. My father had served as an infantry sergeant in the French trenches of World War I. My mother—his bride of a few months—had waited, each day scanning the lengthening casualty lists during the summer of 1918. Years later, each evening in my teens, as we listened to more news of defeats, there were times when I could sense their apprehension.

Yes, it was December 7, 1941. As I walked those few final blocks to my dorm in the fading light of that late Sunday afternoon, the impact of that news on me and my future began to settle in. I was certain that I would be entering the military, though I was not yet draft age. If I had developed thoughts of what I was going to do and where I was going to do it, any conceivable plan for education and career went out the window. I managed to get a phone call through to my parents, and they urged me not to rush into anything.

The semester ended with Christmas break, a time at home when we were at war and the news was universally terrible. I returned to Western and enrolled for my second semester. The best thing about that was meeting my future bride, Helen Freitag—during a class on rhetoric—and having one date. She was majoring in zoology and earned a bachelor's degree in that field, as well as a certificate in secretarial science. The latter served her quite well during my military years. Pearl Harbor foretold, to some degree, our fate but not our ultimate destiny.

I spent the summer of 1942 as a cabin leader at the YMCA camp on Torch Lake. I enrolled for the fall semester at Western soon after I turned nineteen. A few days later, the draft age was reduced to eighteen, so I was expecting to be drafted at almost any time. However, the Army soon announced that young men already in college could enlist in the Army Reserve with the understanding that the Army could call them to active duty at its convenience.

You're in the Army Now

That worked very well for Helen and me. It meant that we had about four months that fall and winter when we dated frequently, most often at Friday night dances on campus. We were seriously in love by the time the Army called me to active duty on April 2, 1943. I was inducted at Camp Grant, Illinois, and assigned to take basic training in the Medical Corps. While there, I was exposed to fellows from all walks of life and standards of behavior that I never knew existed. Those twelve weeks taught me that many American men did not live by the standards that I had previously encountered in my hometowns and college experience.

First, I met a group of conscientious objectors, many of them ministers or lay ministers, who were older than I and who struggled to do the physical tasks of the training. I greatly respected them. At the other end of the spectrum were men classified as illiterate, many of them from Pennsylvania coal-mining country. Still another group had been assembled in one place so that they could be conveniently treated for venereal diseases. A miscellaneous group seemed most devoted to getting drunk on weekends. Finally, we had some college students like me.

I can't do justice to describing life during this training. Living in a tent with seven other men I had just met and had little to nothing in common with was certainly a challenge. We were also enduring uncomfortable weather—the cold and damp of spring in northern Illinois with its snow and mud. Our days were generally consumed with following orders, including close order drills that I failed to understand, then and now. However, on a few weekends, I managed to get to Kalamazoo and visit with Helen at her parents' home.

After I completed basic training, the Army sent me and approximately three hundred thousand others back to various colleges. They had three programs in Army Specialized Training—engineering, medicine, and foreign languages.[3] I was assigned to engineering, but didn't know where my buddies and I were going as we boarded a train and went to Chicago. Somewhere in Indiana, we turned north and headed toward Michigan. How could I be so fortunate?

I spent the next nine months taking basic courses in that field at Michigan State College. In the process, I earned nine credits that were applied to my post-service college work. It was also fortunate because Helen was only ninety miles away, living with her parents and continuing her education at Western, so we enjoyed several weekend visits. By the time the Army sent me to my next assignment, we were engaged.

I was assigned to the U.S. Army Signal Corps and sent for additional training at Camp Crowder in southwestern Missouri. The Army tests I had taken led someone to conclude that I had a great degree of mechanical aptitude, an assessment that I seriously questioned. Nonetheless, that led to my assignment for training in the installation, maintenance, and repair of the technology of the time—teletype machines and switchboards. Upon completing that course in September, I was shipped to Philadelphia for further training, then given twelve days of furlough.

Helen and I took that opportunity to have our wedding on October 2, 1944. After ten wonderful days of honeymoon in a comfortable cottage on Lake Bellaire, I returned to my post in Pennsylvania to find overseas orders on my bunk. This was the beginning of fifteen long months of separation from Helen and the rest of my family.

It took seven days by train to reach San Francisco. During that trip, I saw for the first time our country's vast open spaces and the mountains—snow-packed peaks and narrow canyons with rushing streams. The old army fort on Angel Island in San Francisco Bay was our temporary headquarters. There we received tropical uniforms and numerous inoculations to protect us from diseases we might encounter in the South Pacific.

Our overseas transport vessel was the USS *General M.L. Hersey*, named after an outstanding Army officer. It was the largest ship I had ever boarded; it was transporting approximately three thousand of us. We crossed San Francisco Bay in its famous fog and passed under the Golden Gate Bridge. As I looked back at the Golden Gate, all my life lay on the other side of the bridge—all the people I knew and, most of all, my loving bride Helen. We knew next to nothing about what lay ahead, but a mutual thought went through us all. When would we see that bridge again? This thought found expression in a slogan that we hoped would prove true—Golden Gate in '48.[4]

It was there that I met Jim Pryor of Grand Rapids, Michigan. We played checkers and chess to pass the time. We endured a few days of mass seasickness, then sailed on the calm, vast Pacific Ocean as we crossed the equator and the international dateline. Two or three friends and I slept mostly on the steel top deck, not very comfortable to be sure, but much preferable to our crowded, smelly quarters five levels below. With my gas mask for a pillow and my rain slicker for a blanket, I managed pretty well. The completely darkened ship was a marvelous place to gaze at the sky of the Southern Hemisphere, dominated by the very prominent constellation known as the Southern Cross. The only land we saw was Guadalcanal, visible on our right as we passed it.

South Pacific

Finally, we made it to Hollandia, near the western end of New Guinea, a major transfer hub for the war's Pacific Theater. The harbor had many ships of all descriptions. We climbed down a cargo net to waiting landing craft. After a quick trip, we were deposited without ceremony on an empty beach, with the Army nowhere in sight. There I began the most unpleasant and most difficult few months of my military service.

The sergeant of our platoon announced that he would try to contact some authority and disappeared towards what sounded like a road. We stood for the next five hours with nothing to eat and just a little water in our canteens.

Then a truck appeared, and the driver said he would take us to a "replacement" center—a temporary tent quarters for troops in transit. It was situated near the airport and Lake Sentani. There we received word that they didn't have enough tents for us at the 3367 Signal Battalion headquarters, so we would stay at these temporary quarters until more room was available.

The master sergeant needed labor for a variety of tasks, so each day he made assignments. We mostly loaded or unloaded trucks. We also assembled goods, materials, and equipment at one location, then transported them to the airport, unloaded, and reloaded for waiting transport planes. The Army was moving north to new landings on the island of Leyte. It was so busy that the center's mess hall stayed open twenty-four hours a day to accommodate the traffic flying in and out of that airport. Some troops were "walking wounded," returning from battlefronts with injuries that weren't too serious—a bandage around their head or an arm in a sling. Others had tops of their shoes cut away, exposing sores on their feet.

Their toes were purple from treatment with potassium permanganate for what was called "jungle rot." If you didn't get your feet dry at night, these sores would form and refuse to heal. Other traffic was reinforcement troops moving forward toward battle zones.

One night I was assigned to guard duty. At midnight, a sergeant drove me to a remote post surrounded on three sides by huge trees of virgin jungle. In the middle of this clearing—possibly seventy yards on each side—was a water tank, and it was my duty to guard it. The driver left with his Jeep, saying, "Your replacement will be here at 2 a.m." It was an assignment that I often recall, thinking, "what was this kid from Bellaire, Michigan, doing in New Guinea in the middle of the night, leaning against a huge jungle tree and guarding, with just a loaded rifle, a water tower that the Japanese might want to destroy?" Nothing happened, but I will never forget those long, scary hours.

After a few more days, we got word that the battalion headquarters now had room for us. It was much more pleasant there because our tent was erected over a wooden floor. We were in a grove of coconut trees planted in the square, obviously an organized plantation in peacetime. A small stream was nearby, providing water for adequate showers and the laundry. We had some work to do, but lots of freedom, too. So, Jim and I hitchhiked up and down the fifteen or twenty miles of available road, observing native people. Needy families were also walking the road, the men carrying spears and the women always behind them, usually carrying a child and some other burden.

One day, three of us walked up the side of the mountain, made our way through a cave and up through an entrance to a higher level. There we saw a fifty-foot waterfall cascading over a cliff to a beautiful pool of clear water. It looked to me like it ought to have trout in it, but of course it didn't. There we enjoyed some tins of field rations and a few cans of beer, unfortunately without the alluring presence of movie star Dorothy Lamour dressed in a sarong.

After a few days of comfort in tents with a wooden floor, twenty of us were assigned to build a warehouse near the shores of Hollandia Harbor. We soon discovered that this project was poorly planned. A 200′ × 200′ site, about ten feet above the harbor level, had been cleared and leveled, and we had a supply of building materials, including concrete and a mixer. We found a tent large enough for about fifty men to put up their cots and mosquito netting, although our only latrine was a slit trench. We had no shower in this location—just one degree from the equator, where the temperature was often near 100 degrees, with lots of rain and mud.

In the tent where most of the men and the officer slept the first night were also several piles of plywood sheets. My friend Jim and I looked at each other, then walked down the trail through the jungle to the next encampment, where we "borrowed" two canvas cots. Thus, we could set up our mosquito nets. From that single action, we acquired a reputation as "scroungers."

From our slightly elevated location, we could see quite a lot of lumber floating along the lush mangrove-lined shoreline, some of which came in handy for the construction project. One man in our group—named Wakefield, like the town in the western Upper Peninsula of Michigan—was the only one who had any idea about how to proceed with construction. He told Jim and me that he would make tables for us and build a latrine if we brought him the lumber. He also helped us build a raft to use in extracting some of the "found" pieces of wood from the roots of the mangroves.

As Jim and I paddled around the bay looking for lumber, for the first time in my life I saw the array of beautiful tropical fishes that are part of shores and reefs in the South Pacific. The brook trout streams and the bluegill lakes of Michigan contained no such beauty. All the colors were there, fishes of blue and yellow, black-and-white stripes in several shapes and sizes. A new dimension of my interest in fish and fishing opened.

Also on the plus side, we found that the building housing the kitchen was staffed by two of the best cooks in the Army. The day before Christmas, they said, "If you guys can find some pie pans, we'll make cherry pie for Christmas dinner." Jim and I took off for the next encampment, found their cooks, and asked, very politely, if they had some extra pie pans. "Eureka!" They had three to spare. True to their word, our cooks made some wonderful cherry pies and helped make our holiday a little merry. Enjoying this delicious pie is the one thing I remember about that holiday in 1944. It was a delightful contrast with the monotony of uninteresting preserved foods that we ate a lot.

Battling with Mosquitoes

A day or two after Christmas, a special messenger came with orders for me to return to our headquarters encampment for a teletype-technician refresher course. Our function was to build and maintain communication systems for the Air Force, which, during World War II, was still part of the Army. I settled into an empty space in a comfortable tent and was informed that I had been promoted to Private First Class (PFC).

During the two or three days that I spent in the course, I learned that I could replace worn-out electrical contact points with pieces of coin cut to the proper shape. It was best to use Australian coins because they had a higher silver content and, therefore, would conduct electricity better. I also learned that if you need to string wire high enough to keep it off the ground, never try to put it in a tree that has an ant's nest.

Unfortunately, within a couple of days I realized that I was running a high fever. First, I reported to sick call, where the fever was verified. Then I was transported to the nearest hospital, where I was diagnosed with and treated for dengue fever. I don't remember much of the first three days while the fever raged, but I was in a large ward with as many as fifty other beds. This mosquito-borne disease also had a very apt nickname—breakbone fever.

Jim came to visit and told me of developments back at camp. It seems that Wakefield and others had constructed a latrine (like outhouses in rural America), largely out of the lumber we had salvaged from the harbor mangroves. A second officer assigned to the project arrived. A day later, a sign appeared on the structure—"Officers Only." The latrine mysteriously caught fire and burned down the following night. Perhaps only those who have crouched over a slit trench would fully appreciate the animosity between enlisted men and officers that this incident created.

Soon after my ten days in the hospital, I was assigned to a twelve-man team. The war had moved north to the Philippines. Our assignment was to set up weather stations at airstrips being established to the west and north. We had men of several technical backgrounds, and I was trained to install, repair, and maintain the teletype machines.

We flew westward on a C47 two-engine transport plane the two hundred miles to the island of Biak, where we built and equipped a weather station. My task was easy—put in the wiring, build a suitable location for the teletype machine, and check to make sure it was fully functional.

After a long, boring, inactive wait, we flew to the island of Leyte on a C54, which had four engines and a much longer cruising range. While waiting there in temporary quarters, we caught a glimpse of two or three nights of Japanese air raids, mostly to the east of us, searchlights and anti-aircraft fire in the distance. They weren't that close to us, but these were my first sights and sounds of warfare.

Armed and ready.

HOWARD A. TANNER COLLECTION

Instead of building a weather station on Leyte, we moved on to the island of Mindoro—headquarters of the 310th Bomb Wing—with four airstrips, a squadron of heavy bombers, a fighter group, and a squadron of A20s (a low-level fighter bomber).[5] There we found quarters in a transient camp and quickly assembled the materials, constructed, and equipped a weather station at the Maguire airstrip.

Then our team was divided. Two sections moved northward, and I was left alone to maintain and repair the teletype equipment at four airstrips, a fifth one at the local naval base for PT boats, and at the headquarters building. In this assignment, I finally found the tasks where I could apply those things that I had been trained to do, but nothing happened for nearly a week. I stayed by myself in a nearly empty transit camp, awaiting orders and figuring that the system had lost track of me. It was, perhaps, the loneliest time of my life. I remember the very large rats running through the tent that I shared with no one else.

One night I enjoyed talking with a group of American men who had been trapped in Manila at the sudden outbreak of the war. Our advancing troops had just liberated them from a Japanese prison camp, where they had been held since the war began. They had received little or no information from the outside world during their incarceration and were hungry for news—of any kind.

Then one day a fella named Charles Marchiando arrived in a Jeep, saying, "I'm looking for a guy named Howard Tanner. I have orders for him." He told me that the Jeep would be my vehicle to function as a one-man detachment. I didn't see Marchiando again until 1967, when he wrote me a letter about the salmon fishing explosion, wondering if I were the same Howard Tanner he had worked with on Mindoro. We renewed our relationship and visited often until he died.

Mine was a very small piece of the war in the Pacific. The information that came through on my machines was used by observers to draw the maps that the weather forecasters would use in briefing the bomber and fighter pilots flying each day. I lived and made friends with the enlisted men assigned as weather observers. Soon I was promoted to technician fifth grade (equivalent to a corporal) and the following month to technician fourth grade (equivalent to sergeant). I'm proud that I kept the machines running, except when wires broke or when water soaked to the point where nothing operated. Then I drove the Jeep from strip to strip and especially to the headquarters to deliver the information that was critical to the pilots flying combat missions. As the war moved northward, the number of machines I serviced dwindled, I had fewer duties and responsibilities, and I was left in what I might call a backwater of the war.

As I most often worked in solitude, mail from home was extremely important to me. Helen wrote a letter each day that I was overseas, and my parents wrote about once a week. Unfortunately, my frequent movements often interrupted their delivery; I once waited thirty-eight days for my mail to catch up with me. These letters were the thread that connected us to our wives, to our families—a connection to the real world. Mail call and receiving letters were the high points of each day.

Sick of War

It was early August 1945 when I was diagnosed with hepatitis and hospitalized. One day the mimeographed hospital newspaper reported that we had dropped a bomb on Hiroshima that was one thousand times more powerful than any previous one. I thought that someone had added too many zeros. Soon we dropped one on Nagasaki, too. Then came word that the Japanese had surrendered and the war was over.

When I was discharged from the hospital, I resumed my duties, which had diminished considerably as airstrips shut down and the 310th Bomb Wing head-quarters moved north. Sometime in early October, the tents I shared with my weather-observer colleagues were destroyed by the edge of a monsoon. Fortunately, we negotiated with the Navy and occupied a very comfortable Quonset hut.

With the war officially over and with little work to do, I sought opportunities to explore and to hunt and fish. The old expression "you can't keep a squirrel on the ground in tree country" was appropriate. During this time, I became friends with a Filipino family. Damaso and I fished together with cast nets, and his wife did my laundry.

Of course, I looked forward to going home, and the Army had a very fair "point" system for scheduling who returned home and in what order. The longer you had served, the more points you had. Those with service in combat areas and those with wounds received additional points. Each month, those eligible for heading home were announced. I waited my turn and, based on my points calculations, figured that I would be eligible to go home in January or February 1946.

A Very Close Call

Around the first of December, having checked the teletype machine in the weather center after breakfast, I departed with Damaso and one of my friends for some fishing and a look at new territory farther upriver. I returned late that afternoon to a very dramatic turn of events. After I had departed in the morning, a radio message came through, urgently ordering me and another man in our group to "take the first available aircraft to Manila." By then, we had just one "courier" flight per day that would get us to Manila. The other fellow had followed the order, but died when the plane crashed in the South China Sea and lost all aboard. Then another radio message came in, ordering me with even greater urgency to take the next available flight, which would be the following morning. I went to say goodbye to Damaso, pick up my laundry, and gather my rifle and toolkit.

The next flight got me safely to Manila, where I found my orders to return home, some letters, and Christmas packages. Although I had not seen him, I had filed regular reports to my commanding officer Lieutenant Hunt over the previous year. Then I encountered him while walking towards his office on base. Before I could salute, he offered me his hand and, with tears in his eyes, said, "I thought we had lost you."

After a few days waiting and being processed and renewing friendships, thirty-five of us boarded an old cargo ship. We journeyed across Manila Bay with Corregidor on our left and the tip of the Bataan Peninsula on our right. It was December 7, 1946—five years to the day after the bombing of Pearl Harbor.

Home for Good

The trip home seemed to take forever. The ship encountered the edge of a typhoon, and we made a stop at Guam. With 750 troops on board, meals were difficult. The kitchens bolted to the deck were unable to operate in rough weather, and it took us thirty-one days to arrive at Honolulu.

It was so wonderful to see this beautiful piece of America with normal activities, American movies, and wonderful food. In a few days, I boarded a very comfortable converted passenger ship, which took just seven days to sail under the Golden Gate and arrive in San Francisco. After two or three days of processing, it was a fast train trip to Chicago, then more paperwork, new uniforms, and finally a pass to downtown, where I knew that Helen was waiting.

Checking in at the hotel desk, I learned that Helen had left a message that she was out having dinner with her father, who had arranged our hotel reservations. I waited just a few minutes in the lobby before she came through the door. I followed her to the elevator, where she came into my arms. It was the single happiest moment of my life, and what a way to celebrate her birthday—January 25, 1946! My war was over; I was discharged the next day. Helen and I had four or five wonderful days together, then took a train to Kalamazoo, where both sets of parents greeted us.

My military service was over, but the war stayed with me as I returned home to settle down. I realized just how fortunate I was, especially since I had no wounds or significant injuries, and considering my experiences with tropical diseases and the close call I had by missing the original order to fly to Manila. My assignment required that I use all that I had been taught about communications technology and function independently within a team. My exposure to a very different geography, ecology, and climate helped me appreciate the wider world. My interaction with folks of a foreign culture taught me respect for differences and possibilities for collaboration. It also stimulated my curiosity and encouraged my openness to try new approaches in various circumstances. My relationship with comrades enlarged my horizons and increased my vision of what might be possible. My observations on fishing and tropical fishes gave me greater insight into what would soon become my professional field. My time in service was such a sharp contrast to what I had experienced up to that point. I believe it added so many dimensions, shaped my understanding of people and places, and so very significantly changed what I understood, what actions I would take, and how well I might work with people.

Like 17 million others, I had served my country, had fortunate training and assignments, and arrived home safely, although I had seen and heard men die. I had never imagined being in the military, much less near the frontlines of a world war. However, I gained greatly from the experience and was better prepared for the next stages of my life, especially better than I would have been had I spent those four years in a quiet town or an orderly college campus. To come home to the provisions our country made for returning veterans in the GI Bill was a real reward for my time served.

Building the Educational Framework

Exactly what lay ahead I couldn't tell, of course, but Helen had written me a letter nearly every day. Before we married, "What do you want to do (after the war)?" had been the focus of numerous discussions. I would tell her, in a very unorganized way, that I wanted to do something related to fisheries. During these exchanges of letters, she undertook to explore which colleges and universities had the best programs that would prepare me for a career in the developing field of fisheries management. She looked at both freshwater and saltwater programs. She sent me the curricula of colleges and universities in California and Florida, as well as the University of Michigan, Michigan State College, Ohio State University, and Cornell University.

By the time I headed back to the United States, we had decided that I would enroll in the fisheries program at Michigan State College (MSC). Clearly, Helen did the necessary research to change my undefined, undirected general thoughts of a career in fisheries to a specific course of action that would lead me there. So, in the spring of 1946, I enrolled at MSC to pursue a degree in zoology, with a specialty in fisheries.

MSC—Bachelor's

I was somewhat familiar with Michigan State because of my time there in the military. Still, having forgotten my pre-college visit there, I had to search to find the zoology department, which was then located in Morrill Hall. The fish lab, such as it was, was in the basement. My first task was to sit in front of Dr. Hunt, chairman of the department, while he reviewed the mixture of college credits that I wanted to transfer to Michigan State. Fortunately, most of my credits from Western were acceptable; I also had credits from the Army Specialized Training Program, as well as a class or two of so-called "military educational effort" provided after the war was over to keep us busy. I also had courses in economics and philosophy.

Dr. Hunt was very precise and polite, but demanding. He finally decided that I had a term and a year to go to finish an undergraduate degree in zoology. This meant that I was in my junior year; however, I hadn't taken a lot of zoology. So, for the next term and the next year, most of my courses were in this subject. Almost all had a laboratory component, too, so I took a lot of hours and had a lot of labs. Eventually, I took all the zoology that was available. I also took histology and histological techniques, as well as genetics, mammalogy, animal ecology, ornithology, ichthyology, and limnology. I think I got all A's.

One of the reasons I did so well in school then was that my personal life had changed so much. I had gone from being a disoriented eighteen-year-old freshman, who really didn't know what he wanted to do and was expecting to go to war soon, to a twenty-two-year-old man who was much more serious about things. Having been in the military, I had some difficult living behind me, and now my life with Helen was coming together. We were both focused on a specific goal, and I suppose it's harder to explain a poor grade to a working wife than it is to a distant mother. For the first year or so, Helen worked part-time for a minister who needed a secretary for some administrative duties. She had completed her business course, and she was a skillful and experienced secretary. We were happy, well-situated, and had an idea of where we were going. Therefore, the good grades followed.

Then, I also met the professors who were to be so critical in my education. Chief among these was Dr. Robert Ball, who had arrived in the fall of 1946 to teach limnology. I also made the acquaintance of Dr. Peter Tack, who was also important to my professional development. Life was busy and sometimes difficult, but we felt very fortunate all the same.

Financially, we had our savings, a small salary that Helen was earning, and $90 a month from the GI Bill, which also covered books and other materials. The GI Bill even provided me with a pair of hip boots when I took ichthyology. Along the way, we acquired a dog, a camera, a sewing machine, and our first automobile—a 1931 Model A Ford.

As I think back, the vast majority of veterans who entered college after World War II were starting at the freshman level. I was two or three years ahead of most of them, so my classes weren't quite so crowded. Also, when it came time to find a job, my advanced standing meant that I would be ahead of the pack. As I have often remarked, I have been fortunate with the timing of certain key events in my life.

While fisheries had always been fascinating to me, I found limnology—the study of lakes and streams—to be just as interesting, if not more so. Dr. Tack taught ichthyology and a course called aquaculture. The latter course consisted of a sequence of studies of various aspects of fisheries, from commercial fishing to fish hatcheries. It made up the bulk of my studies in what I would call fisheries management.

Looking ahead to graduating in the spring of 1947, I started applying for jobs. I had tentatively decided not to go to graduate school, though both Dr. Tack and Dr. Ball had vaguely suggested that it might be a good idea for me to consider furthering my education. But I wasn't sure that I wanted to spend more time in school.

MSC—Master's

One day, however, Dr. Ball called Mercer Patriarch and me into his office and said that he had two fellowships from the Michigan Conservation Department. We were the two students from the senior class that he thought were most eligible, and he offered us each a fellowship for a master's program. Dr. Ball was conducting a series of experiments in fish hatchery ponds, and if I accepted the fellowship, I would be working at a warm water lake in the Indian River vicinity. The experiments involved looking at the effects of commercial fertilizers on warm water lakes, trout lakes, and subsequently on trout streams. I excitedly discussed the offer with Helen, and it was obvious that it was too good an opportunity to pass up.

Dr. Ball was my major professor for both my MS and PhD degrees. He came from Ohio, did his undergraduate work at Ohio State, and his graduate work at the University of Michigan under the famous limnologist Dr. Paul Welch, who had written our textbooks on the subject.[1] He also gained valuable experience by working at the Institute for Fisheries Research (IFR) while he was a graduate student. Dr. Ball had just started teaching, and I was in his first limnology class. When he awarded the fellowships to Mercer and me in the early spring of 1947, we became his first two graduate students, and I would become his first PhD student in the summer of 1948.

Besides serving as my major professor, Dr. Ball became my mentor, editor, and friend. He set high performance expectations, teaching me how to write papers and to conform to the standards of science. He also taught me all the methods for collecting and analyzing the field data that would be the basis of my thesis and dissertation. Bob remained at MSU for his whole career, and he served as major professor to more than one hundred graduate students, becoming by far the most productive faculty member of the new Department of Fisheries and Wildlife in that regard.

Later, when I was MSU's director of natural resources, Bob and I worked together repeatedly. By that time, in addition to his role as professor of fisheries and wildlife, he had become director of MSU's Institute of Water Research (IWR). He saw to it that I was appointed assistant director of IWR, and together we developed the concepts behind "third stage" treatment of wastewater and built the treatment systems for the campus water plan.

My master's research examined the effects of adding inorganic fertilizer to a natural warm water lake. The practice of adding fertilizer to aquatic systems to stimulate the growth of fish populations was a hot subject at that time. Dr. Homer Scott Swingle at the University of Alabama was a pioneer in this area.[2] He had fertilized small ponds in fish hatcheries and farm ponds and had produced spectacular increases in the growth rate of some varieties of sunfish. After that, researchers began to focus on the possibilities for duplicating his results in other contexts.

Dr. Ball had been selected by the Michigan Conservation Department, through the auspices of the IFR at the University of Michigan (UM), to carry out some studies. He began by assigning me to a project on a warm water lake—North Twin Lake—in Cheboygan County. This twenty-seven-acre lake, now called Cochrane, was located due east of Indian River, and was entirely state-owned. This was the first time I was paid by the Michigan Department of Conservation.

Dr. Ball worked with me to develop an outline of my field activities for the summer of 1947. The plan was for me to work half-time on my project and half-time

Helen's ready for fishing.
HOWARD A. TANNER COLLECTION

with District Fisheries biologist Walt Crowe, who was a friend of Ball's from their days at the IFR.[3] Dr. Ball had a great deal of respect for Crowe's abilities as a field biologist, and I soon acquired the same respect. He was excellent.

Guided by Dr. Ball, I established a weekly schedule of data collection. As we applied the inorganic nitrogen, phosphorus, and potash fertilizer (NPK) each week, I ran a series of chemical measurements of water quality, sampled and recorded water temperature, took plankton samples, and analyzed the stomach contents of fish that I caught.[4] One very time-consuming activity was the collection of bottom samples.

Helen and I enjoyed our life together on the lake, although the conditions were a bit primitive. Dr. Ball had also arranged for a state employee with a pickup truck to haul our trailer from East Lansing and park it near the shore, in the forest of medium-sized jack pine. All our potable water had to be hauled in twenty-gallon cans from a faucet outside the railroad depot in Indian River. We bathed in the lake, or in the Pigeon River about half a mile away. I shaved every morning from a

Our Kozy Coach home and truck.
HOWARD A. TANNER COLLECTION

basin placed on a small platform nailed to a tree and used a little metal hanging mirror. Our toilet and garbage disposal was a pit that I dug about twenty yards from the trailer. I nailed two 2 × 4s to trees so that they spanned the pit and attached a toilet seat in the middle. We had a Coleman lantern for light and an icebox for food. Our doors and windows were screened so that we could escape the evening mosquitoes, and Helen made strawberry shortcake using wild strawberries. My experience of living on Pacific islands during World War II certainly came in handy when it came to establishing the elements necessary for comfortable living apart from "traditional" facilities.

We worked hard, with Helen helping me in many ways. Most significantly, she kept records and labeled all my collections. Dr. Ball came to check on my progress at least once a month, and I worked with Walt Crowe on his summer assignment with the sea lamprey assessment project, directed by Dr. Vernon Applegate. Part of this work was to search for the presence of sea lamprey in all the Great Lakes tributaries of the "tip of the mitt"—from Charlevoix on the west to Cheboygan on the east.

When I worked with Walt Crowe, we followed a very particular procedure. I would approach a pair of sea lamprey on their spawning bed, in an area of a stream that had a clean gravel bottom. With a small frog spear, I would spear both the male and female, and we would place them in a fruit jar of preservative that we properly labeled as to date and location. Working with him that summer provided me with many valuable learning experiences. When I assumed my duties as Michigan's chief of fisheries in 1964, I made sure that Walt moved to Lansing to be on my staff.

When the season ended, Helen and I returned to East Lansing, and I began the slow process of writing up my results from the lake fertilization experiment. In summary, we concluded that inorganic fertilizer applied at certain rates to a warm water lake in Michigan could be expected to increase the size and number of fish present.

However, the next winter, Dr. Ball, Walt Crowe, and I visited North Twin Lake to take water samples through the ice and found that something disconcerting had happened: not enough oxygen was left in the lake to sustain life. When a lake freezes over, no more dissolved oxygen gets into the water through wind and wave agitation. When snow covers the layer of ice atop the lake, the aquatic plants cannot conduct photosynthesis—a process that uses carbon dioxide and releases oxygen.

The bacterial decay of organic matter continues in the winter, and these bacteria use up oxygen, so that the amount of dissolved oxygen declines until the spring thaw renews the sources of oxygen. If the oxygen declines too far, "winterkill" occurs. What had happened in North Twin in 1948, which had never been known to win-terkill before, was that the increased organic matter produced by fertilization had decayed and used up all the supplies of dissolved oxygen in the lake. All vertebrate and invertebrate life died, rendering North Twin Lake dead. The conclusion that fertilizer produced the winterkill was inescapable.

The situation was a bit like the old saying "The operation was successful, but the patient died." We found that fertilizer could produce more and faster-growing fish. However, a lake in Northern Michigan is not the same as the ponds Dr. Swingle had worked on in Alabama. Our early conclusions also included a growing conviction that fertilization of lakes would probably not be cost-effective. We had shown that fertilization had caused increased growth, and that it probably increased production in terms of pounds of fish per acre. But Michigan has more than ten thousand lakes. Also, the winterkill showed clearly that we needed to do more research on how much fertilizer could be added without producing such devastating consequences.

I presented my preliminary report that winter at the Tri-State Fishery Conference, held at the training-school center on Higgins Lake (now known as the Ralph A. MacMullan Center).

MSC—PhD

Dr. Ball and the research staff of the IFR were interested in taking the next step, and they developed a project to further explore the relevant questions about fertilization. I participated in the planning for this as a junior researcher, knowing I would be offered a fellowship to pursue my doctorate. I did not even finish writing up my master's thesis at the time, since we plunged into the next project so quickly. In fact, I did not complete my MS thesis until 1950.[5]

Certain tentative conclusions guided us in our project design. We knew that we needed to find out how much fertilizer could be safely applied, and we decided that fertilizing warm water lakes, of which there were thousands, would not be effective. However, only a small fraction of Michigan's inland lakes have temperatures that are cool enough to be suitable for trout. Further, trout populations in the inland lakes generally need to be maintained by stocking, making the management of trout lakes already a more expensive proposition. So, our general conclusion was that if there were an acceptable role for the practice of artificial lake fertilization, it would be in the management of trout lakes.

In the Pigeon River State Forest, twelve to fifteen miles due east of Vanderbilt, lie six small state-owned trout lakes. Five of them had steep shorelines, making them difficult to approach for someone carrying sampling gear, rubber boats, and fertilizer. Plans came together quite quickly. I started fieldwork on those lakes during the summer of 1948, and we applied fertilizer, at varying rates, to five of the lakes. The sixth was left unfertilized as a control. All the rates of application were less than what we had employed on the lake that I had studied for my master's project. Thus began a project that would take almost five years, culminating in the artificial overturn of one lake in the summer of 1952.

In early summer of 1948, Helen and I drove a conservation department pickup truck to Vanderbilt, from which we went over fourteen miles of rough gravel road to a site we had chosen for our trailer. It was a beautiful, remote spot about one mile downstream from the cluster of log buildings that were, at that time, the headquarters for the Pigeon River State Forest. Our site was on a bluff overlooking the river and a forested campground. Helen and I spent three enjoyable summers on that site, joined in the summer of 1950 by our son Mark, who had been born in December 1949.

The first summer was spent establishing baseline data. This meant that I had to measure, sample, and describe all the characteristics of each lake before adding fertilizer. These measurements included the details of water chemistry: hardness, pH levels, dissolved oxygen levels, and temperature patterns of stratification. I also did depth readings with a Secchi disk, as well as calculating invertebrate abundance and distribution of aquatic vegetation.

I sampled each lake at least once every week and determined their fish populations. If a lake is to be managed effectively for trout, all other species of fish need to be removed, because trout are very poor competitors and will not thrive when other species are present. I found that four of the lakes had other species, such as yellow perch, bluegill, green sunfish, pumpkin seed sunfish, white sucker, common shiner, and a few brook trout. So, it was necessary to treat them near the end of the summer in 1948 with the piscicide rotenone to remove all potentially competitive fish species.

The lakes were stocked with brown trout, and in 1949 and 1950, we applied fertilizer to five of them. I then measured the same factors each summer: changes in temperature patterns, chemical factors, dissolved oxygen, fish growth, bottom fauna abundance, and a few others. The work was hard. The sampling schedule demanded that I do it "rain or shine," and in August of 1950 we had seventeen straight days of rain. We finally loaded all of our wet clothes and diapers for our son in the car and went to "the jail" at my parents' home in Antrim County (where my father was still sheriff) to dry out.

The first year, I had only a rubber boat that I carried from lake to lake in our Model A Ford. My father had helped me to modify this old car by cutting out the rumble seat and replacing it with a small truck bed. It was just right for hauling my gear. By the next year, I had acquired, from various places within the conservation department, five old and very leaky rowboats. I spent a lot of time caulking and painting each one. I placed one of these boats on each of the five lakes that had very steep banks, and I used the rubber boat only on Hemlock Lake, where the launching was easy.

In the evenings, we had work to do. In bad weather, we worked inside the trailer by lantern light. In good weather, we worked near the campfire. Helen kept two sets of our data records, mostly on three-by-five cards. We tried to keep these two sets of records in different places, since the data were irreplaceable and there were no such things as computers or storage disks in those days. At night, I also studied French and German, since I was required to pass exams in two foreign languages for the PhD degree. For the things that I was interested in, the study of foreign languages was a huge waste of time, and they are no longer required for natural resources doctoral students.

The summer of 1950 was our last field season on the trout lakes. At the end of the project, we again poisoned the lakes to remove and collect all the fish. These fish provided the last, critically important measurements of the effectiveness of adding fertilizer. An extra crew, including Dr. Ball, was there to help with the collection.

Unfortunately, I was not there. As part of our initial collection process, I had been catching fish on hook and line and releasing them after clipping a fin for identification. I stuck one finger with a fish hook, it got infected, and I had to go all the way to Gaylord to see a doctor. He gave me a shot of penicillin, and we scheduled two more shots several days apart. I made it in for the second one, but, since the finger looked great and I was pressed for time, I didn't return for the third. The infection returned, and I did the same thing. When I finally went back to the doctor, he was very unhappy with me and he gave me, I suspect, a larger than normal dose. We were about to poison the lakes, and I needed to be there, but I started to swell up with giant hives and it quickly got worse.

I was in pretty bad shape, so we left the supper dishes on the table and headed for Bellaire, with Helen driving. When we got there, they put me to bed, and the doctor stayed with me all night. At about 5 a.m. he decided to drive me to the hospital in Petoskey. I don't remember the ride or the next couple of days, but I recovered after three days and left the hospital. The next day I returned to the lakes and found that all the fish had been collected and most of the work was done. I felt quite guilty for not having been there, but Dr. Roger later told us that I had had a very close call. I never took penicillin again.

The field data collection was over. Three summers' worth of samples and data now waited to be shaped into a thesis, and I still had more classes to take. Since I was going to be receiving all my degrees from the same department of the same school, I didn't have any other classes in my major area, with the exception of advanced genetics. Yet I still had to accumulate more credits. So, I took a lot of botany classes, including one on taxonomy of aquatic plants and algae. For one project, I had to identify a large collection of stream insect samples taken by someone who had been an instructor in a Civilian Conservation Corps (CCC) camp in the 1930s. I also took two courses in statistics, plus public-health courses in bacteriology and in the engineering features of sewage collection and treatment systems. These courses proved helpful later in my career.

Dr. Ball and I made a winter trip in early 1951 to check on oxygen levels in the six lakes, and I made one last trip later that winter. These were difficult expeditions; we got to five lakes by using snowshoes. Cutting holes through the ice, taking water samples, then analyzing them for dissolved oxygen at temperatures well below freezing was a challenging task. But the results were good, in that while we found some oxygen depletion, none of the lakes was threatened with winterkill.

The important conclusion that I reached following the completion of two graduate degrees relating to the artificial fertilization of natural lakes can be stated very briefly. I think that fertilizers should never, under any circumstances, be added to natural freshwater lakes. But perhaps some explanation is in order.[6]

Glacial Lakes

A natural lake begins as a basin filled with water. In North America, a very large percentage have been formed by glaciers. Lakes start with very little in the way of nutrients for plant and algae growth. Over long periods of time, nutrients are added by water runoff from surrounding soils and inlet streams, and by leaf fall from terrestrial systems. Nutrients accumulate, normally over hundreds or thousands of years depending on how nutrient-rich the soils in the watershed are. The lake then moves in stages from being full of clear water with no rooted aquatic plants to having a rich stand of aquatic plants and the water rendered less clear by planktonic algae. The basin bottom gradually accumulates deposits of organic matter. Thus, the lake will become shallower, and emergent aquatic plants, such as water lilies and cattails, will encroach. Eventually the lake becomes a marsh, until it finally forms solid ground, which we might call a muck farm.

Anyone who deliberately adds nutrients, or otherwise speeds up the natural process, is moving the lake towards extinction by a quantum leap. Therefore, to do this robs future generations of the enjoyment that a lake brings. In my opinion, then, it is unethical and should never be done.

It was my major professor's concept and plan to expose me to, and acquaint me with, all aspects of aquatic environments—water temperature, life cycles of invertebrates, stratification, flora and fauna, water chemistry (oxygenation)—the environment that fishes depend on, seasonal cycles, and optimum factors in predator-prey relationships. He expected me to articulate my findings in preliminary and final form, with presentations in writing and public speaking.

Over my six years in graduate school, I took courses in a wide variety of fields, including botany, statistics, entomology, bacteriology, and public health. I served as a graduate assistant in a laboratory program, and I was occasionally called upon to give a lecture. My fieldwork encompassed almost every aspect of data collection and observation in my academic area. I also had to ultimately conform to the high standards for academic writing that Dr. Ball insisted upon.

Ancillary to my actual degree program, I had the opportunity to meet many professionals in the field of fisheries, to attend professional meetings, and to present papers. The contacts that I made on campus and within the ranks of the Michigan Conservation Department served me well in many unexpected ways in the future.

Eventually I earned three degrees, including the PhD, at Michigan State, with much support during my graduate work coming from the Michigan Conservation Department. It took me ten years to complete my formal education, but it prepared me well for a career in my chosen field.

Postdoc Project

I had reason to expect to stay in Michigan once I had completed my graduate studies, but that didn't happen. The opportunity I did have that summer was to work with Dr. Ball and Dr. Frank Hooper of the University of Michigan—both outstanding limnologists. They wanted to discover what could be learned through the artificial overturn of a lake in midsummer. My PhD dissertation on lake fertilization, combined with my working knowledge of the unique trout lakes of the Pigeon River country, made me a good candidate to help in one of their research projects. I was interested in finding out whether an overturn could mimic some of the effects of commercial fertilizers in increasing a lake's fish capacity.

I earned a salary on the project while I waited for another job opportunity to develop, and it also provided a valuable opportunity to collaborate with these two esteemed researchers. West Lost Lake in Otsego County—one of the six lakes I had studied during my dissertation research—was chosen for the overturn.

During the summer, most deep lakes in the northern United States, including the Great Lakes, stratify into three layers or zones. The upper zone, known technically as the epilimnion, will usually extend from the surface to a depth of about twenty feet. This is the biologically productive part of the lake. Wind and wave action keeps these waters oxygenated, and the water temperatures remain fairly uniform as they are warmed by the sunlight. These waters, and the lake bottoms associated with them, are where most of the invertebrate animals—crayfish, insect larvae, snails, clams, etc.—live and reproduce. Rooted aquatic plants are limited to this zone by the depth of sunlight penetration, and free-floating algae and animal plankton also exist primarily in this upper zone.

Below the epilimnion is a layer known as the thermocline. Generally, it is a narrow zone of four to eight feet and is much lower in productivity due to colder temperatures and a lack of sunlight. It is, however, cold enough to provide habitat for fish such as trout and salmon, which cannot tolerate the warmer epilimnion. During the summer, it also acts as a buffer zone between the upper and lower regions of lakes, so that the mixing action caused by wind and waves never reaches the depths. This colder water also acts as a temporary barrier to algae and zooplankton that are dying or disabled and are in the process of sinking into the next layer—the hypolimnion—and to the lake bottom. Fish that feed on zooplankton find an abundance of food in the thermocline.

Throughout the summer months, this lower layer (the hypolimnion) accumulates large amounts of plant and animal matter that sink there from the productive layers above. This lower layer is often without oxygen because oxygen is consumed by the decaying process of the material arriving from above. Generally, sunlight does not reach to these depths, and life in this area is limited to those organisms that can tolerate a lack of oxygen and the much colder temperatures.

These stratified lakes, with the naturally accumulating nutrient material in the unproductive lower level (hypolimnion), overturn naturally as cold weather comes in the fall. Falling air temperatures cool the upper waters and, as the water cools, it becomes heavier and sinks to the bottom. When the water temperatures approach uniformity from top to bottom, the thermocline breaks up. Then, wind action circulates the entire lake volume, not just the epilimnion, and all the nutrient-rich waters (which provide the fuel for biological growth) become available for the abundant plants and animals of the upper layer. Unfortunately, however, this process always happens in the late fall when it is too late for optimum biological activity.

Our research team had a good understanding of this entirely natural process, but the question arose as to what would happen if we could produce an overturn in midsummer when sunlight and warmth could combine with the newly available nutrients that we would be pulling back into the upper levels of the lake. There, in warmer waters and sunlight, these nutrients would be available fuel for biological growth processes.

We began the project by running two lines of irrigation pipe from a generator and a large electrically driven pump on shore. One pipeline went to the center of the lake's bottom. The other line, suspended by a series of barrels, ran to the center of the lake's surface. The pump sucked the water from the bottom (hypolimnion) of the lake and discharged it as a spray on the lake's top center. Then we sampled and analyzed what happened when this cold, nutrient-rich water devoid of oxygen was distributed into the sunlit warm waters of the lake's surface.

Physical and biological responses occurred. Physically, we eliminated the whole volume of the hypolimnion, restoring it to the water area above the thermocline. Biologically, both phytoplankton and zooplankton increased immediately. What we had postulated was proven to be correct; in other words, nutrient materials reintroduced into warmer sunlit waters produced an immediate jump in biological activity. These results were published in the *Journal of the American Society of Limnology and Oceanography*, after I had moved to Colorado following the fieldwork. I had only reviewed a draft in the publication process.[7]

This was rather basic research, and we had little or no expectation of direct application. Still, it added significantly to our understanding of the dynamics of biology in freshwater lakes. By creating an artificial overturn, we were able to document more precisely what the overturn meant in terms of returning biological nutrients to the natural food-producing systems of the lake. Furthermore, by this time it was becoming quite clear that the addition of fertilizers to natural aquatic systems was not a wise or practical means of increasing lake productivity. Perhaps, though, by mimicking the annual overturn of a lake we had found that increased productivity could be achieved by merely altering the timing of natural events.

To make the experiment work, we had to keep this generator and pump going without interruption for five or six days. That meant that the fuel and oil had to be checked about every six hours. One morning as I arrived to service the generator, I saw a fisherman sitting on the bank nearby. After I had serviced the machinery, I walked over to greet him. As he sat there studying our setup and smoking his pipe, he said to me, "Can you tell me what you are doing?"

I said, "Well, we are pumping the water from the bottom of the lake and putting it on top."

He got up, looked at me, and started to walk away.

"That's what I thought," he said.

And over the years I've often wondered what he said to others as he explained what he had seen that morning on West Lost Lake.

Professional Practitioner

I n June 1952 I finished my PhD. I was the one who officially received the degree, but Helen had worked just as hard as I did during those years. Since June of 1947, we had been partly supported by scholarships from the Michigan Conservation Department IFR. Helen and I had, understandably, expected that I would be offered employment there when I finished my degree, but no offer came. And, although I had numerous feelers out for a job, I did not have anything "in hand" when I graduated. So, I worked temporarily on the overturn project.

Colorado Here We Come

In late spring, I began corresponding with Dr. William Beckman, leader of the Colorado Cooperative Fishery Research Unit at Colorado Agricultural and Mechanical College (now Colorado State University).[1] Dr. Beckman was a graduate of the University of Michigan and was a member of the IFR there when I met him a few times during my early years in graduate school. He had helped me read the scales of the fish samples that I took during my master's degree field studies on North Twin Lake in Cheboygan County. I was negotiating for an opening as an assistant professor in the Department of Forest Recreation, chaired by Professor Jack Wagar, in the School of Forestry. I had sent out applications to various other institutes/agencies/organizations, but this one seemed to be the best opportunity and the most likely.

Around the first of August, the job was offered to me, but it was two thousand miles away. This made the decision difficult, especially since Helen and I were only children whose parents were still in Michigan. We had never given serious thought to moving so far from home. Helen had never been west of southwestern Missouri, where she had visited me once during one of my stateside Army posts in 1944. I had only seen "the West" from a troop train on my way to San Francisco, en route to overseas duty in the South Pacific during World War II. But it seemed that we really had no other option available, so we took it. I was to report to Fort Collins in early September, in time for fall classes.

After saying tearful goodbyes to our parents, we bade farewell to Michigan. Our few pieces of furniture were shipped, and we had two baby beds tied to the top of our 1950 four-door Chevrolet. Bright and early one morning, we crammed all our clothing and two small sons into the back seat and took to the road.

It was a very long, hot trip across the cornfields of Illinois and Iowa, then the plains of Nebraska and Wyoming. There the landscape looked barren to us compared to the lush trees and farms of Michigan, and we were lucky to see a few cows once in a while. Several times we just looked at each other, silently wondering what we had gotten ourselves into. At those times of near despair, we would get out our letter from Professor Wagar, in which he described Fort Collins as a small town of eighteen thousand people "lying in the lush irrigated valley of the Cache La Poudre River and nestled next to the foothills of the Rocky Mountains." That would give us a little bit of hope.

At Wellington, Colorado, only twelve miles north of Fort Collins, we entered a dramatically different landscape. Suddenly we saw lush, irrigated farms, scattered cottonwood trees, the Rockies with their snow-capped peaks, and the town of Fort Collins among the low ridges of the foothills in the foreground. Finally, we had reached our destination.

We received friendly greetings from Dr. and Mrs. Beckman and Professor Wagar and his wife. We even stayed temporarily in a back bedroom at the Beckmans' with the boys in two cribs and Helen and I sleeping in a twin bed! Soon we found a house to rent near the college. It wasn't new and it wasn't big, but it looked great to us. Rent was $90 a month, and we acquired the kitchen stove by paying another

$10 a month. We had a house with three bedrooms, central heat, and a bathroom. We also had a yard with a garden, a small garage, and a cherry tree. Our life as a faculty family had begun.

Academic Leadership

The Colorado Cooperative Fishery Research Unit, hereafter referred to as the fish unit, consisted of a three-way agreement in its purposes and functions. The U.S. Fish and Wildlife Service provided the leader and the leader's salary; the state fish and wildlife department provided an operating budget; and the college provided the housing for the unit and integrated the instructional program.

The unit's principal function was to educate young men (there being essentially no women in this subject area at that time) to become professional fishery biologists. The subject matter of the graduate student research was selected jointly by the three elements of the unit. It began with the understanding that these research projects would concentrate on fishery problems in warm water reservoirs.

I was to be assistant leader of the fish unit under Dr. Beckman and would teach all the undergraduate fishery classes in the department. My first year's salary was only $4,250, but all I cared about then was that I finally had a position.

Colorado Characteristics

Getting to know the territory was essential to my success in this situation. In the middle of World War II, I had crossed Colorado on a train, so I knew that there would be vast areas of open plains, mountain canyons, and high peaks. But it's one thing to pass through as a traveler and quite another thing to go there to live and work.

First, Colorado is so very different from Michigan. The center of the state has more than fifty mountain peaks that are more than fourteen thousand feet above sea level.[2] Amidst these peaks are the state's only natural lakes, many rushing streams, lovely lush valleys, and unequaled scenic views.

Eastern Colorado, by contrast, is dominated by high, dry plains where irrigation makes it possible to grow a variety of crops, sometimes with record yields. This area—from the foothills of the Rockies to the eastern borders with Nebraska, Kansas, and Oklahoma—is one vast plain with, basically, no natural lakes. Within a twenty-mile-wide strip running north and south immediately east of the Rocky Mountain foothills is a string of cities from Cheyenne, Wyoming, southward to the New Mexico border.[3] This is also where the vast majority of Colorado citizens live—1.5 million people then and 5.3 million now.[4] The reservoirs there, most of them lying close to the Rockies' Front Range, are filled annually with snowmelt from the mountains.

The western third of the state—west of the Rockies—is a land of mesas, canyons, and a near desert climate. It also has, essentially, no lakes, but is crossed by the Colorado, Green, and White rivers and some other small streams.

All of this covers a huge area—more than twice the land area of Michigan. This landscape itself would present any fishery biologist with an array of lakes and streams far different from, and far more diverse than, those of my native state.[5]

Second, the history of the area influenced a very different approach to resource management. Water scarcity was entirely new and alien to anything I had ever known. In Colorado, water was a commodity, and nearly all surface waters belonged to somebody—some water company, some irrigation company, some farmer, some rancher, or, occasionally, some city. Just as miners who had taken the risk of "prospecting" would stake a "claim" to precious mineral resources that they discovered, so, too, would individuals, businesses, and governments claim control over scarce, essential water resources. Western water law was based on the premise of "prior appropriation," or first come, first served.[6] That presented significant additional problems for anyone engaged in fisheries management because this approach seldom, if ever, considered the needs of other creatures that also depend on water.

The streams were gushing torrents in the spring, fed by melting snow and the release of water stored for irrigation and domestic purposes in high mountain reservoirs. The U.S. Bureau of Reclamation was in the process of building an immense system of diversion from the Colorado River on the Western Slope through tunnels to the large reservoirs—Horsetooth to Carter to Estes Lake.[7] Other reservoirs with similar function were the Cheesman on the Platte and the Pueblo on the Arkansas.

By the late 1950s and early 1960s, Denver was purchasing more water rights and planning to construct delivery systems that would divert considerable additional amounts of water for use by its residents. We in the Colorado Division of Wildlife, as it was called then, tried to prevent the serious damage or total destruction that the proposed expanded diversion for Denver might wreak. Reservoirs, created for irrigation and for drinking water for the major cities, lay scattered across the Eastern

Plains—mostly close to the mountains. These were the "fishery" environments that we were supposed to manage.

I quickly discovered the diverse characteristics of the state's surface water. Snowmelt water was a practically sterile environment with little or no development of the food chain. The Eastern Slope rivers were gushing torrents as the mountain snows melted in the spring months and as waters were released to fill reservoirs on the plains. However, they were practically unfishable in the winter because the water levels might drop twenty vertical feet each year between June and fall. These were definitely not the stable waters of Michigan. They ranged from the unproductive waters recently emerged from glaciers and snowpack to the naturally occurring water on the plains that resembled seawater in its amounts of dissolved salts.

These unstable aquatic environments were a very difficult area for fishes to grow and thrive. They would go nearly dry and sustained little in the way of trout populations. It was equally difficult for a fishery biologist to do anything about it. Typically, the plains' reservoirs had the European carp as their principal fish population. This species could tolerate these vast fluctuations and still survive. Along with the tench, it had been planted in Colorado before 1900.

Furthermore, most of the reservoirs that we were supposed to manage were privately owned. Often no boats were allowed. I once collected samples of the rare Rio Grande cutthroat trout from the tiny streams of an old Spanish land grant on the east side of the San Luis Valley. There they had been protected from public access for a hundred years.

These are the conditions in which we operated. However, even in these generally unfavorable environments were some interesting fish populations. By the late 1950s, fishery managers there had had eighty years of experience with introducing new species. The rainbow trout was the principal hatchery product.

Brown trout were present in streams, and some beautiful streams were scattered throughout the mountain parks, with lower gradient flowing through meadows and rangeland, which belonged almost exclusively to large ranchers.

Brook trout introduction proved immensely destructive to the native cutthroat trout populations. Furthermore, due to the brook trout's slower growth rate, it most often produced only stunted fish of little interest to the angler.[8]

To summarize all of this, the physical nature of the terrain and the genuine shortage or lack of productive environments for fish population were a completely new challenge to me.

Managing

Except for a few natural lakes in the mountains and the mountain streams containing principally trout populations, other waters that the public had available for fishing opportunities consisted principally of the constructed reservoirs. The fish populations there consisted of species that we selected and planted. We were making what we hoped would be sustainable introductions of prey and predator species. Since essentially no native species were available, we did not face—or we did not often face—the question of displacing or disrupting native populations. Since we were dealing with reservoirs that sometimes went dry or were drawn down to a point where fish populations were eliminated, our stocking of fish and the choice of species that we might use was an ongoing form of management. Think of us as people with two buckets—one of which contained our species of choice for stocking and the other filled with rotenone.

The People Factor

The diversity of Colorado's geography, history, water resources, and resource management law were also important pieces of background that my Colorado experiences gave me. However, by far the most important piece was the interaction that I had with the people I worked with—the diversity of problems that we faced, and projects that we undertook.

The characteristics of Colorado citizens were probably the most important aspect of all. I think it reflected their recent history and the attitude of people existing in a very sparsely populated environment. Perhaps I could describe it as self-reliance or the willingness to accept all problems demanding a solution.

Working with the "Greatest Generation"

When I returned to Michigan, the public knew that I had worked in Colorado for several years. Many colleagues and fishery constituents understood that I had become professional acquaintances with West Coast fisheries biologists. They understood that those relationships were an asset in obtaining coho salmon eggs. The public has never, however, learned about the significance of twelve years of professional experience where the introduction of new species was commonplace.

If one examines the pattern of fisheries management during those years in Colorado, a certain understanding emerges. Colorado's fishery division had hired a staff of about a dozen trained and qualified fishery biologists in the years following World War II. With hindsight, I more fully appreciate the working environment I had during my Colorado years (1952–64). Most of the men I worked with would later be described by Tom Brokaw in *The Greatest Generation*. We had grown up during the Great Depression, and most of us were veterans of World War II. For example, several of my coworkers and graduate students had been bomber pilots in Europe and East Asia; others had been submarine officers, tank commanders, paratroopers, or infantry soldiers. In other words, we had all made sacrifices, lived in dangerous areas, and carried out numerous difficult tasks.

In addition to our common experiences of the Depression and World War II, many of my colleagues had unique backgrounds in western culture and environment. Within some families, fathers had been early forest rangers and park rangers. For the sons, growing up on large, sparsely populated properties had given them unique levels of assurance and a wide variety of skills.

Most of us had also completed college under the GI Bill, and we were eager to apply newly acquired scientific principles to the vast array of problems and opportunities presented by our assignments. We were the first generation of wildlife biologists and conservation professionals to have the training to systematically and scientifically understand the biological problems presented by the diverse lakes, streams, and reservoirs that were the habitats of Colorado fish populations. So naturally, we saw it as our mandate to overturn most of the old ways of managing public resources and fish hatcheries.

Although we were not well paid, we did have funds to support our programs. We also had solid leadership and great freedom of action. If we decided to do something, or to try something new, we could generally get the permission and funds to do it. We worked hard, got things done, and found ways to have fun in the process.

It was also a part of our style that when we could perceive that something in the way of fisheries management had to be done, or that there was opportunity to take action that would make things such as fishing better, we didn't hesitate to do it. Sometimes we did the right things, sometimes the wrong. But at least we never held back out of timidity. This explains a background of experience that made it easier for me to make the decision to introduce coho salmon into the Great Lakes. But it is only a partial explanation.

Everything related to resource management, including fishery resource management, was totally, and I mean totally, different in Colorado. What I learned there, the adjustments that I made to the application of my scientific understanding, gave me a totally different mindset, a very different understanding, and quite a different approach to the way we made our management decisions. Working in a vastly different fishery environment gave me a truly different attitude about what was appropriate in the way of good, ethically correct fishery resource management.

Changes and Challenges

Dr. Beckman left for Washington, DC, in 1953, leaving me as the only fisheries professional in Fort Collins with a PhD, and for years the only university-trained fisheries person in Fort Collins. Within six months I was a member of the graduate faculty of Colorado State. I also became leader of the Cooperative Research Unit, where I was called upon for all kinds of assignments.

My graduate students worked on a variety of issues that never have and never could have occurred in the abundant waters, lakes, and streams of Michigan and the Upper Midwest. For the most part, our research might properly be labeled experimental management. As chief of fisheries research, I was called upon to put together programs that would help answer some of the following widely diverse subjects and issues with a scientific approach.

- Could we find a way to reduce the evaporation from reservoirs without biological impairment? Long-chain fatty alcohol would be spread one molecule thick across the surface of reservoirs during periods of high temperatures.
- Would the biology of the lake be damaged? Eighty-five percent of rainbow trout in some private fish hatcheries had liver cancer. We built a research project to help explain this epidemic.
- All the fish in the new lake that we as a department had built were dying. What was the cause?
- How can you sample fish populations that are distributed in different

vertical sections of a deep reservoir. Answer: you build a vertical gill net and use it effectively. Never done in Michigan.
- Electroshock fishing gear to sample stream fish populations doesn't work well in many Colorado areas. Why? Build something that will work. We did.

The fish shocker had always worked well in Michigan. However, in Colorado, it worked in some waters but not others. Why? The answer to that question came soon. The waters recently emerging from masses of snow and ice had no electrolytes, so an electric current would not transmit. On the plains, the opposite was true. As soon as water had spent some time in soils of areas that were almost semi-desert, it acquired so many salts that there was almost a direct short across the electrodes, and again, it was likely that the electric shocker wouldn't work. That was the easy part; what to do about it? We worked with engineering student colleagues and were in the forefront of developing shockers capable of modifying their output to work in a variety of waters with different conductivity. We also worked on simple things such as creel censuses to measure the success or failure of different experimental stocking.

Professionally Speaking

I think I am an effective public speaker, transmitting information and explanation to a variety of audiences, from scientific colleagues to various groups of the public that we serve. There is, in my opinion, no better way to become prepared for these tasks than being required to lecture every day—well, most every day—for several years to an audience intensely interested in the subject matter. In my case, they were junior and senior students, usually taking the classes as part of their major. Most of these students were of greater than average maturity—veterans of both World War II and the Korean War—studying seriously and obtaining an education to prepare them for their chosen field. As a professor, you'd better have something substantive to say and you'd better present it well. I'm confident that I taught many things to my students, but I also know that I learned much from them.

I also found myself speaking to many other audiences, ranging from sportsmen's groups to scientific meetings. I was called upon to lecture for in-service training sessions on lake and stream environments, fish populations, and general management procedures. These audiences included people coming from the National

Park Service, the U.S. Forest Service, and the Bureau of Reclamation. I journeyed to Wyoming and New Mexico to give similar lectures to a variety of personnel from those states' game and fish departments.

I also learned how to and how *not* to perform as an expert witness. An expert witness is allowed to express opinions within the subject area of his/her acumen and established expertise. While I was employed as chief of fisheries research for Colorado, one of our fish hatcheries was sued for water pollution. I testified extensively that there was no measurable pollution in the stream where water from the fish hatchery was discharged. I had extensive data from my personal field collections that proved there was no measurable change brought about by the effluent from the fish hatchery. I testified for several days and several hours to this effect. After I completed my testimony, I explored farther downstream—farther than I had collected any samples. There I found that the turbulent, rapidly flowing stream that I had sampled turned into a black, fetid, stagnant, smelly pool immediately below a cascade. All my testimony was completely wrong. I had failed to be sufficiently thorough in my investigation. It was an honest mistake but terribly wrong. I didn't know what to do. In this case the judge ruled against us and forced us as a state agency to first close the hatchery and then remedy the problem that we had created. He saved my ass. The lesson I learned in that situation was to be very, very sure that you are as thorough as possible before drawing conclusions.

VIP Assignments

Around 1961, when I was selected to be chief of the Fisheries Research Division of Colorado Game, Fish, and Parks Department (CGF&P), I was also the only PhD degree holder in the department. This got me, or I think this is what got me, some special assignments. I had the distinct feeling that my director, Ernie Woodward, used me in situations where he wanted to impress somebody with the fact that he was sending a well-qualified staff member. For example, one Sunday afternoon in midsummer, then governor John A. Love of Colorado was fishing on the Arkansas River. Heavy rains had washed out some mining retention deposits containing cyanide, and the fish population of the river died almost before his eyes. In response to the emergency request from the governor, Director Woodward sent me with the crew to investigate the urgent demand for attention.

In an amusing incident, President Dwight Eisenhower was coming to Colorado for a vacation. He was scheduled to spend his visit on the ranch owned by Robert Six, then owner and president of Continental Airlines. Mr. Six asked for some special attention, and Director Woodward quickly informed him that Dr. Tanner would visit with him immediately. I don't know whether he was impressed or not, but I agreed to plant large numbers of trout in the private waters of Mr. Six's ranch. Everybody understood that the president was supposed to catch lots of trout and spread the word of the wonderful fishing to be had in Colorado.

This kind of background was important when it came to my job as chief of fisheries in Michigan, where I had to explain the salmon action to a variety of audiences, many of which questioned and some of which disagreed with our plan. I would be asked to explain why we were closing the long-standing program of stocking catchable-sized trout in streams. Occasionally, I would be required to defend various departmental programs in legal and corporate arenas.

To Introduce or Not?

During the formal education that followed my World War II experiences, I was thoroughly immersed in a well-established dogma: never introduce an exotic species. However, the history of fish introduction in Colorado was entirely different and produced an entirely different background of understanding in professional fisheries biologists. Let me explain.

As a relief from the rigors of early European American settlement, the people of the cities and towns, farms and ranches of Colorado quite naturally sought out fishing opportunities. Yes, in the mountains were some beautiful trout populations. There were at least three, perhaps four, different races of cutthroat trout—the greenback cutthroat, the Rio Grande cutthroat, the Colorado River cutthroat, and perhaps another race of cutthroat in the Arkansas River headwaters. Beyond that was very little in the way of game or sport fish, except perhaps the somewhat less popular Rocky Mountain whitefish, which was native to the Western Slope.

However, from the eastern borders of the state to the mountains, there were literally no fish. In a similar vein, there were no fish in the lower levels of the Western Slope of Colorado. That was understandable in that no natural lakes existed in those very large areas of the state, and rivers often went dry or nearly so. The European American settlement of Colorado brought with it irrigation systems, and in these systems were many reservoirs. It seemed quite logical for folks to want to fish in those waters, especially since "naturally occurring" waters were so scarce and scattered.

During the same era, Michigan experienced early introductions of various species. Keep in mind that this would have been in the latter years of the nineteenth century and the early years of the twentieth, well in advance of any real science or understanding of the effects of introducing new species. But understand also that because no native fish species lived in these reservoirs, almost any introduction seemed worth a try.

During this period, European (common) carp was distributed by the federal government along with bass, trout, and tench. The carp proved to be a sustained disaster.[9] Many Colorado reservoirs to this day have a fish biota comprised principally of the European carp. The tench exists in only small locales in the San Luis Valley.

However, also around this time, lake trout and grayling were introduced, in addition to the species mentioned earlier.[10] The lake trout remains there, sustained by hatchery stocking in a few reservoirs. During my time in Colorado we had no sustained population of grayling.

To step back a moment, scattered throughout the early introductions were other species as well, mainly the warm-water fishes such as yellow perch, walleye and northern pike, largemouth and smallmouth bass, bluegills and other sunfishes, and numerous species of minnows. My point here is that there was an extensive background of introduced species of fish in the early history of the state, and with some exceptions, these efforts had produced desirable results—fishing opportunities for the public in a great percentage of the available waters.

Beginning in the 1950s and continuing through the years that I was in Colorado, the U.S. Bureau of Reclamation completed a series of mountainous or foothill reservoirs of several thousand acres. So, it was natural that people asked for the development of fish populations in these reservoirs near their homes. The Colorado Game, Fish, and Parks Department had the management responsibility. Some of the most active programs of the department at that time involved, in one way or another, providing additional lake and stream areas for public fishing. If the public has no place to go fishing, and the U.S. Bureau of Reclamation builds a lake, and the state agency stocks it with a fish population that is reasonably self-sustaining, it's pretty clear that you have done something for the public and you will have their support.

When I moved there from Michigan, with my established attitudes concerning the introduction of exotic species as undesirable and irresponsible, I encountered a whole new environment and established attitudes. New fish species were desirable—the whole fishery was based substantially on introduced species. Yes, there had been some undesirable introductions, but those had occurred when fisheries science and understanding of predictable results were lacking. Now we had professionally trained fishery biologists confronted with new challenges and actively searching for new species to add to Colorado's warm waters and to the state's colder reservoirs in the mountains.

Reflecting

I have often reviewed in my mind and in conversation with others the variety of experiences that I had in the general field of fisheries management and research during my twelve years in Colorado. Many aspects of those Colorado experiences would have little relationship to the vastly different geography and water resources within Michigan. The circumstances of our work were so different than what you would be likely to encounter in Michigan. For instance, we did research on a group of mountain lakes, where it took a horseback ride of fourteen miles to get to the site. I don't dislike horses; I just don't like to ride them.

Close Calls

I also had a few accidental close calls. While sampling walleyes on a remote reservoir in the Arkansas Valley on the high plains of southeastern Colorado, one student and I raced our boat across a lake to find refuge before being struck by a dust storm. We made it. We huddled—four of us in the cab of a pickup truck. When the storm struck, our boat rolled over and over down the shoreline, dumping our motor into the water. The blown sand took the paint right off the windward side of the truck's cab.

I had the privilege of working with my students on Trappers Lake at the head of the White River, high in the mountains in a roadless wilderness. On one occasion, we were snowbound for three days in September. In the process of collecting information and experimentation, I was also buried in an avalanche. I flew by helicopter in the dead of winter to sample and find volumes of flow in a wilderness stream. On another occasion, I shot myself, accidentally, with a device designed to poison coyotes with cyanide.

The breadth and diversity of these experiences and others prepared me to accept challenges and to undertake new actions and programs.

Salmon Stocking

The activities of one fishery biologist that I worked with in Colorado were especially pertinent. His name was Dick (W. D.) Klein. A former naval officer, Dick was working with the Colorado Department of Game, Fish, and Parks before I arrived in that agency. In 1961, the department was reorganized, creating two divisions—a division of fisheries management that would remain in Denver, and a division of fisheries research that would be located in Fort Collins with me as the chief.

Dick moved to Fort Collins and was a member of my staff. He was frequently selecting new places to plant kokanee salmon and, on one occasion, coho.[11] As I began to pay more attention to his efforts to introduce salmon, I found his kokanee stories very interesting. The kokanee was potentially an excellent choice then for several reasons. First, several new, large reservoirs were filling up for the first time. Shadow Mountain and Granby were located at the headwaters of the Colorado River. Granby Reservoir was the largest body of water in Colorado at that time, with a total surface area of about six thousand acres. Blue Mesa was under construction on the Gunnison. Horsetooth and Carter Reservoirs on the Eastern Slope were some of the largest waters in the state of Colorado, several thousand acres in size. They would be cold-water reservoirs, suitable for trout and salmon. The reservoirs' handicap was that nearly all would fluctuate in depth due to the diversion of water to cities and farms. This meant that the bottom of these reservoirs would be largely and frequently exposed, thus eliminating much of the food production that occurs on the bottom of natural lakes and some reservoirs.

Second, the kokanee was a tasty sport fish and filtered small zooplankton and phytoplankton from the water column, making it a very likely choice to do well in reservoir environments.[12] Dick found moderate to excellent success in numerous places in Colorado. During my years at the agency, kokanee was firmly established.

In about 1961 or 1962, Dick shared with me his hope of obtaining coho salmon— my first encounter with information about that species.[13] Little did I know or dream that what I learned about coho then would be so important after returning to Michigan. Indeed, at the time I had no plan to return, but my discussions with Dick taught me quite a lot. He had been trying for some time, without success, to get a supply of coho eggs. The principal reason that he encountered was that global populations had declined so drastically—especially in the Columbia and Fraser river drainages—that the states in the region and the Province of British Columbia had signed agreements not to share coho eggs with any location outside those drainages. Perhaps it was very important that we tried hard for a few years to find a source of coho salmon without any success.

Dick and I frequently discussed what the coho would need if we were to introduce them. I was unconvinced that we could find a food source for the species. Their typical food requirement was some species of small fish that would not be present in the Colorado reservoirs that we were attempting to stock. Dick maintained that he thought that the sucker populations of those high mountain reservoirs would be utilized by the coho. It doesn't matter that he was wrong. When we eventually did get some coho, they never grew to a satisfactory length, primarily because they never had a good supply of prey.

Of course, many sources had said that coho would not survive in the fresh waters of Colorado. Dick told me that there were a couple of examples where coho had completed their life cycle in fresh water. For one, the state of California had held coho in their freshwater hatcheries through a full life cycle.[14] Even more important was an example in Montana.[15] Montana had introduced coho salmon, perhaps a decade earlier, into the waters of their Georgetown Reservoir. There the coho grew, matured, and spawned in fresh water. Subsequently, the population was lost, but the biologists had a clear record that the species had completed its life cycle. At that time, three of my former students were working for the Montana department, and checking with them verified the Georgetown project results. That became important for me later as I defended our plans to introduce coho into the Great Lakes.

Dick finally succeeded in getting a limited number of coho eggs—about 150,000—from a private fish farmer in the Northwest. These eggs were hatched in the Colorado Fish Hatchery, and the fish were stocked in Granby Reservoir, where a young biologist named Larry Finnell was assigned to various studies, including following the coho salmon.[16] Shortly before I left Colorado, I had occasion to fish there, principally for trout and kokanee, and was pleasantly surprised to catch one or two coho. They were small (about twelve inches long) and nearing the end of their life cycle. At the time, this information, this source of experience, did not seem important, but it became important when I began to defend the wisdom of stocking coho in the Great Lakes.

In discussions with Dick about his activities and plans for salmon, I learned a great deal. I brought these memories with me to Michigan, where some of them found application and became a **critical factor** in the Great Lakes salmon story.

Colorado Conclusions

I spent my first twelve years of professional employment in Colorado—a couple of years dealing only with students, a few years leading the Cooperative Fishery Research Institute at Colorado State, some years supervising graduate research, and a few as chief of the Fisheries Research Division of the Colorado Game, Fish, and Parks Department. Those responsibilities were rewarding in many ways. The challenges I faced there provided experiences and perspectives that I would never have had in Michigan. My time in Colorado also gave me more than just practical experiences; it gave me a different way of looking at the world that I could not have gained by staying in the Midwest. It shaped, to this day, the way I think, the way I work, and the way I seek solutions to problems.

Those Colorado experiences also had a formative effect on my perspectives as to what fisheries managers could do and what they should not do. Added to my Michigan background and education, my immersion in the types of innovative and exciting programs we established in Colorado made it possible for me to have the vision, competence, and boldness required to proceed with the momentous changes that we ultimately made in the allocation of Great Lakes fishing resources. In summary, my exposure to the variety of problems we encountered in Colorado shaped my perspective and expanded my willingness to consider a broader array of options to "fix" the problems we faced with the Great Lakes when I returned to Michigan.

When the coho salmon program was underway in the Great Lakes, I was asked one question over and over. Sometimes it would be asked by my critics, and sometimes by those who simply wanted to understand the success we had been able to achieve in balancing the fish populations of the Great Lakes while creating new recreational opportunities (along with an economic boom). The question was asked in different ways. In the early days, after we had reared the first salmon eggs, but before any fish had been released, the question might be phrased in a neutral way, like this: "What makes you think that your efforts will succeed, when salmon have been tried many times in the Great Lakes and all of those attempts were complete failures?" Or my worthy opponents might challenge my professional ethics by asking, "Don't you think that it's stupid (dastardly, reckless, etc.) to introduce an exotic species into a system that's so large and shared by other states and Canada?" After the initial spectacular fishing developed, the same question would be asked, but in a more admiring way. They would say, "Where did you ever get the idea to plant salmon?" or "Did you have any idea what it would mean to Michigan, as well as to the other states and Canada?" or "How did you have the courage and audacity to do it?"

I have thought about these various forms of the same question many times, and I believe that the answer lies in the sum of experiences I had in the West. The freewheeling western style of doing things had forever altered my approach to challenges and opportunities. Thus, when the opportunity to plant salmon in the Great Lakes came, I was ready to seize upon it and commit to the boldest decision I ever made.

In the midst of the Michigan salmon-introduction episode, I told Helen that I would become either famous or the biggest bum in the history of fisheries management. But by that time, I wasn't a person who gave away an opportunity through hesitation. When I decided that something was the right thing to do, I just did it. I am sure that I would never have become that sort of person, however, had I never lived and worked in Colorado.

The Great Lakes
Context

The Inland Seas

I was certainly aware of the Great Lakes in my early childhood. I had played a small role in Great Lakes fishery research as a graduate student when I helped assess the spawning habitat for sea lamprey in Northeast Michigan streams that feed Lake Huron. However, I still had much to remember and learn about these amazing lakes as I took the next major steps in my career.

The Great Lakes form the largest surface freshwater system in the world; some have even called them "inland seas." This huge system is not in a distant, remote area, but in the center of tens of millions of people and in the industrial heartland of North America. The Great Lakes were formed and filled over a span of two million years, most recently by retreating and melting glaciers at the end of the last ice age—between nine thousand and eighteen thousand years ago.[1]

To grasp the true significance of the changes wrought by the introduction of Pacific salmon into the Great Lakes, one must understand the very impressive dimensions and importance of the lakes to the United States and Canada. This book is written from a Michigan perspective and from a fishery biologist's point of view, so there will be much of that. However, the overall significance of this freshwater system needs to be understood, at least within the broad outlines of comparison

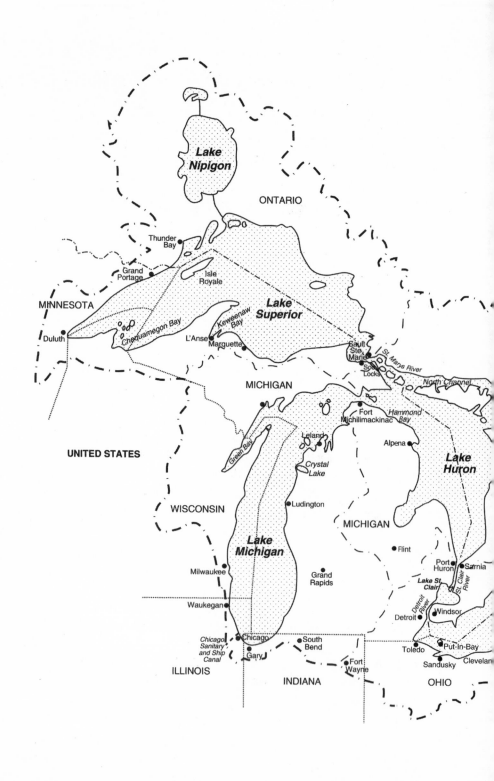

MAP OF
THE GREAT
LAKES BASIN

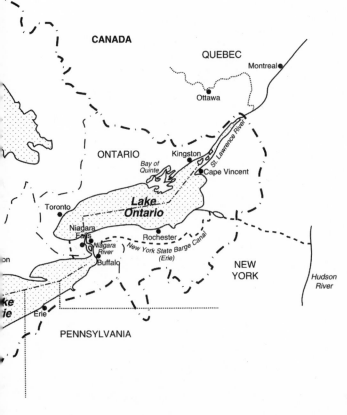

KEY	
··············	State /Provincial Boundary
· — · —	Great Lakes Watershed (Basin)
—·—·—	International Boundary
- - - -	Canal
●	City

CANADA

QUEBEC

Montreal●

Ottawa●

ONTARIO   Kingston

Bay of Quinte

St. Lawrence River

●Cape Vincent

Toronto●

Lake
Ontario

Niagara Falls

Niagara River

Rochester●

New York State Barge Canal
(Erie)

Buffalo●

NEW
YORK

Hudson
River

ke
ie   Erie●

PENNSYLVANIA

The Great Lakes Basin.

MICHIGAN STATE UNIVERSITY EXTENSION

with other fresh waters. Michigan's boundaries encompass 35,000 miles of streams and 38,575 square miles of the Great Lakes.

The surface area of the five Great Lakes and the connecting waters such as Lake St. Clair totals about 100,000 square miles. They span 750 miles from west to east and continue to evolve very slowly as the surrounding land rises. For comparison, if this area were to be transposed, it would cover the combined area of the five New England states and about half the state of New York.[2] If the lakes' collective volume—5,500 cubic miles—were distributed over the entire continental United States, it would reach a depth of 9.5 feet. The total volume is about 20 percent of the world's surface fresh water (only the polar ice caps contain more) and about 84 percent of the U.S. supply.

When these lakes are compared in size to lakes around the world, they are five of the world's fifteen largest lakes. Lake Baikal in southern Siberia is rated as the largest freshwater lake in the world, by volume. Lake Superior is generally listed as the second largest lake in the world (by volume), followed by Lake Huron in the number three spot and Lake Michigan as number four. No matter how you slice it, some of the largest lakes in the world are on all sides of Michigan's two peninsulas.

The strict global definition of a lake is a body of water completely surrounded by land. By that definition, Lake Michigan and Lake Huron are not two lakes, but one, with a narrow portion connecting the two lakes and spanned by a bridge. Furthermore, the water level is the same in both lakes and meets the definition of one lake. Perhaps we should brag about being surrounded by the largest freshwater lake in the world.[3]

Lake Michigan is defined as the largest lake totally within the boundaries of the fifty United States and the largest within any single country. I find it amusing that the second-largest lake completely within the boundaries of one state is Lake Okeechobee in Florida. For comparison purposes, consider that Lake Michigan— some three hundred miles in length from north to south—would be about enough to cover the whole state of Florida.

Size alone is not an adequate source of understanding. These five lakes and their connecting rivers and channels lie in the very center of the industrial heartland of our nation and that of Canada, our neighbor mostly to the north. These vast waters made transportation feasible, and their shorelines became a logical choice for settlement and development of an internationally significant trade and industrial base.

The size of the body of water is important, but the interface—the shoreline, the coastline—is where people enter and use them. The state of Michigan's shoreline alone—3,288 miles—exceeds the shoreline of the eastern United States—from Nova Scotia to the vicinity of Mobile, Alabama. This is where people interact; this is where the benefits of these vast fresh waters enter our society and our economy. And Michigan's Great Lakes shoreline is exceeded only by the Province of Ontario's 5,629 miles, emphasizing the importance to Canada as well.[4]

Moving from the dimensions, think of the multiple uses that these lakes provide. In Michigan, more than half of the people—including those living in Grand Rapids, Saginaw, Flint, and Detroit—take their domestic water supplies directly from the Great Lakes. Other large Great Lakes shoreline cities, such as Milwaukee, Chicago, Toledo, Cleveland, Toronto, and Buffalo, do too. I must point out that, except for Chicago, all these cities also discharge their municipal wastewater into the Great Lakes or one of the tributaries, placing a very important responsibility on maintaining adequate treatment systems throughout the lakes' drainage system.

Also, think of the shipping; think of the products moved between ports on the Great Lakes and the shipping that arrives from points around the globe, now with access to the Great Lakes provided by the St. Lawrence Seaway. Now we know that this is a mixture of benefits and detriments considering that the seaway opened the way for nonindigenous and some notably invasive species to enter the system.

Consider that our agriculture—from vineyards and orchards to field crops and forests, to livestock and truck farms—is dependent upon the weather patterns produced by the moderating effects of these Great Lakes.[5] High-quality Great Lakes water is also important to beverage production and food processing, as well as manufactured goods.

Great Lakes shorelines provide countless hours of leisurely enjoyment, and recreational boating is also important. Of course, there's the fishery, too; more about that soon.

Boundaries and "Ownership"

The boundary line that divides the United States and Canada was finally established soon after the treaty that ended the War of 1812; perhaps it was even quite firmly established after the American Revolution.

Eight states now border the Great Lakes—Minnesota, Wisconsin, Illinois, Indiana, Michigan, Ohio, Pennsylvania, and New York. The Province of Ontario does, too. Because Lake Michigan lies totally on the U.S. side of this boundary line, about 61 percent of the surface area of the Great Lakes lies within the borders of the United States and 39 percent within Canada's. Québec sometimes comes into management arrangements between the two countries concerning these lakes because it is riparian to the St. Lawrence River some distance downstream from Lake Ontario.

Michigan

Geography places Michigan in a leadership role in all matters relating to the Great Lakes. Bordered by four of the five (if you count Michigan and Huron separately), the two peninsulas of Michigan are our central stage. In fact, the boundaries of the state extend to approximately the middle of the upper three lakes, so about 45 percent of the state's geographic area is "underwater."

Speaking of shares of ownership, 41 percent of the Great Lakes' surface lies within the State of Michigan's boundaries, while 20 percent is divided among the seven remaining states. I use this language to emphasize that while Michigan is in a position to share many of the combined benefits of these lakes, it also inherits the preponderance of responsibility for all matters relating to the lakes' health and well-being.[6]

With many decisions being made by our national government in Washington, only the Michigan congressional delegation's constituencies lie almost completely within the Great Lakes Basin and within the subject areas of interest. We are the most affected; we have the most to gain by good policies and the most to lose by bad ones. Through our members of Congress, the state legislature, and many other elected officials, we must continuously be the leaders on the U.S. side of the line. In the past, we have done so; we have led in the environmental cleanup of the lakes from chemical pollutants such as DDT, PCBs, and mercury, etc. We must recognize that we need to continue in this role.

Fisheries

We have also led in the realm of fisheries. It was our initiative that converted Michigan waters of the Great Lakes system from an allocation for commercial fishing to a truly significant sport fishing role, and all of that is a part of this story.

In the early twenty-first century, Michigan licensed the greatest number of U.S. anglers—649,639—and these folks spent nearly 11 million days fishing.[7]

Today, approximately 180 species of fish inhabit the Great Lakes, and a healthy Great Lakes freshwater habitat has been a **critical factor** for their survival and success. However, only a few dozen species are caught in large numbers. In the early twenty-first century, the most popular sport species in U.S. waters (by number of anglers and days spent angling) were bass, walleye, yellow perch, salmon, lake trout, northern pike, steelhead, and muskellunge. Anglers in Canadian waters most often caught yellow perch, bass, panfish, walleye, smelt, and pike. Lake whitefish dominated the dollar value in the commercial harvest, followed by yellow perch, walleye, white bass, rainbow smelt, and chubs.[8]

By the late twentieth century, fish species had been accidentally and intentionally introduced into the system over the preceding 150 years. All these species play a role in the ecosystem. The Pacific salmon that we introduced some fifty years ago has certainly become a significant predator, if not the dominant species, in the Great Lakes food web.

Human History
and the Great Lakes Fishery

Michigan is a very special place, and we have an amazing fishery. What in our history made this happen? I'm not a historian, and I won't try to document in detail the developments that I've learned about over the past seventy years from various sources, but I think you'll find the general story very interesting.

History

Principally because the Great Lakes have been in existence for such a short geological period, the number of fish species native to these waters is very small. Four of the five lakes, except some of their bays and estuaries, were deep, clear, and relatively low in production of food material. Numbers of species and pounds of fish per acre were limited. At the top of the food chain, the predatory species consisted principally of lake trout and burbot, which explains the vulnerability of the lakes' biology: when the lake trout suffered, the whole system was disturbed.

Native Americans were certainly aware of and dependent on the Great Lakes and inland-lake fisheries of the region. However, as early European explorers began to arrive on the shores of North America, they hoped to find gold, such as the

Spanish had found in Mexico and in South America. They found little or no gold. Copper ornaments, including decorative metal breastplates, were not uncommon among the various tribes on the Atlantic coast. When they asked where the copper came from, the explorers received only vague answers—that they were obtained in trade from tribes to the west and north.

In the mid-1600s, the French paddled trade canoes upward through Lakes Huron and Michigan, and onward to Lake Superior via the St. Mary's River. Voyageurs paddling westward along the north shore of what is now Michigan's Upper Peninsula were the first Europeans to discover the source of those copper ornaments traded among the tribes of eastern North America.

Along the shore and, historically more important, on the archipelago that we call Isle Royale, they found primitive pits where early people had discovered and mined native copper—copper that could be fashioned into ornaments without any smelting process. Reports describing the location of these deposits of native copper reached the king of France and his court. Remember that.

More than a century passed. Our ancestors fought England and won our independence, and the peace treaty was negotiated and signed at Versailles, France. Benjamin Franklin, who had been our ambassador to France during most of the years of our revolution, had heard accounts of this native copper and its locations. When it came time to draw the boundary between what is now Canada and our new nation, he made sure that Isle Royale was on the U.S. side of that line.

The Revolutionary War and the War of 1812 opened the Midwest to settlement, including floods of immigrants from the East Coast of our country and new arrivals from Europe. This wave of human occupation occurred in the United States and in Canada, too. The strength of the Native American and First Nation tribes in the area was greatly reduced by their losses in those wars because they had been allies of the British. Our European ancestors began to move into what was temporarily referred to as the Northwest Territory.

Prior to the arrival of these immigrants, fish harvesting by native peoples had little impact on the stability and well-being of Great Lakes fish populations. These people simply were too few and possessed proportionately little harvest capability.[1] However, the tide of European American pioneers and homesteaders that moved north and west after the wars brought change in many forms. They cleared the forests for their settlements. They began to cultivate the soils. They used waterpower to saw logs and grind grains. These actions modified the streams, making them warmer, adding sediments, and diminishing flow at critical times. Each dam blocked spawning runs of fish.

Those who tried fishing also had much greater capability to harvest. First, sails were replaced by steam power. Wooden hulls were replaced by metal ones, and fragile nets were replaced by ones made with more durable materials. As transportation got better, markets expanded, and more folks fished longer and deeper and more effectively. Populations of native fish began to be affected.

The first impact was on Lake Ontario—the only lake where Atlantic salmon were present. Although poorly documented, the Atlantic salmon population in that easternmost Great Lake may have been one of the finest in the whole world.[2] This fish was very vulnerable during its spawning migrations up the tributary streams. It was also available to the relatively primitive open-water fishing capability at that time. Improved gill nets were very effective in the lake itself, and when spawning runs occurred, salmon were heavily harvested. The tribes smoked and pickled the fish.

As settlements developed, dams were built on most streams to provide power to the grist mills to grind the corn and grain and to operate early sawmills. Once in place, of course, a dam completely denied the traditional spawning areas to the salmon. The whole landscape was modified as the land was cleared for immigrant settlement and crops. The stream temperatures were modified so that the habitat was substantially changed. By 1850, the world's greatest Atlantic salmon fishery had been destroyed, and the species was extirpated or nearly so in Lake Ontario.[3]

This was the pattern of settlement and impact on resources as the waves of people moved westward. The lakes themselves were, of course, also avenues of movement and transport. Immigrants came up the canals from the Mohawk and Hudson rivers. Larger boats carried them westward to the ports of Toledo, Detroit, and later Saginaw.

Try to imagine the perception that these people must have had of this vast, open territory. Except for a few prairie areas, the Northwest Territory was heavily forested, and those people were seeking timber from the abundant forests. In the process, they cleared the land. What we term "market hunting" was prevalent, not just for fishes but for all other wild game species. Remnants of the fur trade still existed.

People had little or no concept of resource limits. Natural resources were so abundant that they were regarded as truly inexhaustible. If you wanted more land, you moved westward. If you wanted more timber and more lumber, you could find it in the west and north. In a similar way, more fishes were to be had as you moved westward. There was no such thing as fisheries science and little understanding of fishes' life histories, except what was needed to be able to intercept them during periods of abundance.

As a result, some species were so vulnerable that these impacts meant extirpation or extinction. For example, the habitat of the relic population of grayling in Michigan was destroyed by log drives. The species is now gone, perhaps forever, unless the Michigan DNR succeeds in reestablishing it. Other human impacts occurred. People considered the lake sturgeon undesirable because it was so large that it destroyed fishing nets, so it was harvested and killed whenever possible. It was also misjudged as a poor food item and discarded. Efforts were made to exterminate the sturgeon and very nearly succeeded.

Keep in mind that the fishes to be caught were a public resource to be harvested by those with the capability. As more people came, markets expanded. The transport of fish to market became easier; railroads were developed. Fish could be iced and marketed on the East Coast, so the demand wasn't simply for subsistence catching, but also for market harvest.[4]

By the early 1830s, Ohio had become a state, and Michigan's territorial government moved to the hamlet of Lansing in the 1840s. Much of this westward movement had been accomplished by boat, coming up the Hudson, and crossing to Lake Erie via the newly constructed Erie Canal. Pioneers penetrated the interior areas of Ohio, Indiana, and Michigan seeking sites where they could homestead on fertile land. My ancestors on my mother's side made their way to Michigan and homesteaded in Allegan and Van Buren Counties before 1830. My grandfather Tanner was born in Ohio and homesteaded in Allegan County in 1858.

War for the Lakes

Harbors on western Lake Erie, Lake St. Clair, and Lake Huron became very important. A strip of land along the northern boundary between the state of Ohio and the territory of Michigan that included the Port of Toledo came into dispute. It is often called the Toledo War, and it may have been a war, but according to what I have read, no casualties occurred. The dispute was referred to the United States Congress, and the result was predictable when you realize that Ohio was a state with senators and members of Congress, while Michigan was still a territory. Toledo and the connecting strip of land were awarded to Ohio.

Eventually, the territory of Michigan came to include the rest of the Upper Peninsula and Isle Royale. In what I have read, the leaders of Michigan Territory at the time considered it a poor bargain. However, in fact, Michigan thus gained a major portion of the Great Lakes.

Because of these early historical incidents, the Michigan Department of Natural Resources—specifically its Fisheries Division—has management authority over 41 percent of the surface area of the Great Lakes. The Province of Ontario has management authority over approximately 39 percent. These two management entities are therefore preponderant sources of directions taken in the management of the Great Lakes fishery.

The story is repetitious. The frontier ethic, or rather lack of one, judged all resources to be inexhaustible, and some were harvested until they became scarce. But all you needed to do was move westward. The impact of more fishermen fishing with more efficient gear, bigger and tougher boats, and catering to an ever-increasing demand for fish along the East Coast resulted in serious depletion of the more valuable species long before 1900. Early efforts to regulate seldom succeeded. Facts, figures, and solutions were scarce, and fishery managers lacked basic data and understanding.

States were formed, laws developed, and by 1870, there was growing concern that fish populations were being overharvested. Signs of depletion were present in numerous areas. The effect of the expression "the tragedy of the commons" was quite clear. The response in this state was to create the Michigan Fish Commission—precursor of today's Fisheries Division in the Department of Natural Resources—and to require that fish catch records be kept.

Fisheries professionals began to understand how to regulate the harvest of fish species without depletion, but for decades they lacked the political strength to impose those regulations on the fishing industry. When the Fish Commission attempted to restrict the harvest of fish in danger of depletion, it was whipped soundly by commercial fishing interests in the Michigan legislature. They were financially spanked by having their budget cut by 50 percent in 1897 and told not to attempt to regulate the fishery again.[5]

Introducing Species

And now we come to early deliberate introductions in the Great Lakes. About this time, the knowledge of how to rear trout and salmon was developing in Europe. As this knowledge spread through the expanding portions of the United States, fish hatcheries were built, many of them in Michigan.

Lake trout and whitefish were reared and released by the millions in various parts of the Great Lakes. Unfortunately, most of these fish were too small to survive. It is interesting to note that Pacific salmon were being brought from the West Coast to be stocked in the Great Lakes. Early biologists later documented that salmon had been introduced into one or more of the Great Lakes more than thirty-five times prior to our action in 1966. Few of these salmon produced a viable population; a few individual fish records tell of their survival in scattered locations.[6]

However, in the 1870s, rainbow trout from the West Coast and brown trout from Europe were planted and became established throughout much of the Upper Great Lakes as desirable fish populations, although principally in inland lakes and streams.[7] Deliberate introductions of other species began. We generally consider the brook trout to be a native species. Records of that time are not adequate, but the best picture that I can develop is that brook trout were native only to Michigan's Upper Peninsula, with perhaps a small population in the Jordan River of the Lower Peninsula. Thus, the range and distribution of the species expanded.

In the rapidly expanding fish hatchery development of the 1870s, word came of the European carp, known there as the "Queen of the Waters." It was also introduced not just in the Great Lakes area, but probably in nearly every state and watershed of our nation and Canada. It was, for a variety of reasons, highly regarded in Europe, yet it proved to be a nuisance wherever its populations were established in this country, including the Great Lakes.

The Smelt Story

Then in about 1913, other species were purposely added to the Great Lakes. The introduction of the smelt is an especially interesting story. An early sportsmen's club near Beulah and Frankfort decided that they wanted to introduce Atlantic salmon into Crystal Lake, a large inland lake with an outlet to Lake Michigan. As the story goes, they corresponded with people in Maine, perhaps their counterparts as sportsmen or perhaps including an early fish division of that state, asking for Atlantic salmon eggs. Somewhere in this correspondence, people in Maine advised that if they were going to introduce the Atlantic salmon, they should also introduce their most important prey species—the smelt. The Michigan sportsmen asked that the smelt or smelt eggs be sent as well, and that's what happened.[8]

As far as I can tell, those Atlantic salmon were never seen again in Crystal Lake—what you would expect if they were planted as very early fry. Two or three years later, a man going to the small nearby stream in Beulah for water observed the stream jammed with small fishes—the smelt. From that point on, smelt populations expanded enormously. Rainbow smelt, as they came to be called, were a commercial species in most places of the upper three lakes within ten or fifteen years. There were famous smelt-dipping locations; the area of the "singing bridge" on the Rogue River near East Tawas is probably the most well-known. People came there from all over the state to dip smelt by the bushel.

Some of my earliest memories of living in Mancelona were the spring smelt runs at the mouth of the Jordan River in the town of East Jordan. It was organized pandemonium as fishermen were restrained from the river until full darkness. When the conservation officer fired his gun at nine o'clock, fishing folk moved en masse to the banks of the river. They dipped smelt in simple hand nets. This scene was repeated in most of Michigan's streams tributary to the Great Lakes.

The smelt introduction at Crystal Lake is reasonably well documented. There may have been other introductions into the upper lakes that I don't know about or were essentially undocumented. In any event, the smelt populations expanded to enormous abundance in an amazingly brief time.

People tend to overlook the importance of these small fishes, probably because of their size. However, a careful look at their teeth reveals that they are predaceous. They occupied many of the same areas as the lake trout, including those areas of the lake where lake trout spawn. There is no doubt in my mind, and in the opinion of many other biologists, that predation by smelt effectively reduced the survival rates of lake trout by consuming the young sac fry and emerging fry in their spawning areas.

This extremely rapid expansion of abundance is also an early example of what frequently happens when a species occupies a new environment. It is typically true that, in such instances, the species in question leaves behind many of the factors that limited its population in its native waters. Without these limiting factors in the new area, its expansion is rapid and oftentimes enormously successful, at least temporarily. The rainbow smelt is a documented example.

Today we are faced with a multitude of invasive species. This invasion began in the early 1800s when canals were built, allowing the passage of fish around Niagara Falls through the Welland Canal on the Canadian side, and the Erie Canal connecting the Hudson River drainage and the Atlantic Ocean on the American side. Changes that began on Lake Ontario moved with the tide of human populations up through the chain of the Great Lakes until all were impacted. No doubt one of the earliest and most disruptive invasive species—the infamous sea lamprey—entered the upper four lakes through one or more of these canals in the nineteenth century.[9]

The U.S. and Canadian governments responded to complaints from fishermen and established offices and laboratories of the U.S. Bureau of Commercial Fisheries in Ann Arbor, Michigan, and its counterpart in Canada. The beginning of fisheries science grew in a few centers. One of the principal areas was Ann Arbor, with the bureau and with the faculty of the University of Michigan. More about that in following chapters.[10]

Fisheries Management

For many reasons, it is unfortunate that people learned how to rear a variety of fish species under artificial conditions before we had a science of fisheries biology, let alone fisheries management.[1] Fisheries biology is the scientific discipline that studies the biology of various fish species, as well as the ecology and dynamics of their populations. Fisheries management is taking actions that affect a resource and its exploitation with a view to achieve certain objectives, such as maximizing the production of that resource, all these being **critical factors** for the viability of a fishery. Management techniques include, for example, fishery regulations such as catch quotas or closed seasons.

Fish species were spread all over the world long before we understood the complex interrelationships that exist in an assemblage of fishes in each environment. Widely divergent species were reared in the Orient and later in Europe long before the capability of transporting live eggs or live fishes across the oceans to North America existed.

Hatcheries for several species of trout, others for whitefish, still others for the common carp, as well as little-known species such as the tench, had existed for decades in Europe. In the early 1870s, individual states, the U.S. federal government, and private organizations began to build hatcheries and to expand them across the continental United States.[2] Small hatcheries were developed all over Michigan, many of them by the U.S. Fish Commission (eventually named the U.S. Bureau of Commercial Fisheries).

Some of those widely distributed species effectively changed the way we think about trout fishing. An early fish hatchery on a tributary to the Sacramento River in California is the origin of our rainbow trout, sometimes known as steelhead. These were planted throughout the northern states of this country in the 1870s. Brown trout from Europe arrived in our early hatcheries and were widely distributed again in the 1870s. Many people have forgotten or never knew this and consider those species to be beloved natives. They are more properly labeled "naturalized" species.

Research

In the large sequence of things, fish culture and fish hatcheries developed hundreds, possibly thousands, of years before people understood the critical elements of fisheries management. They didn't know fishes' life histories; they didn't know their environmental requirements; they didn't know the interrelationships of the biological food web; and they didn't know how to teach about them. They lacked the basic tools of how to understand fish populations and how to manage them.

Beginning in the later 1920s, a few universities began to develop the first elements of fishery science. New understandings of the life histories of freshwater fish, of the habitat characteristics required by various fish species, and of the interactions between fish and the elements of the food web were creating a new body of scientific literature.[3] Along with a study of fisheries was the field of limnology—the study of lakes and streams. That, too, could not be defined as an established science until at least the mid-nineteenth century.

Much of this development of fisheries and limnology as a science occurred in the northern Midwest at several universities—Michigan, Michigan State, Wisconsin, Ohio State, and Cornell. Canadian universities—the University of Guelph, for example—were also involved. This was the region of enlightened education if you look at the fisheries profession and the names that were prominent in those years. The majority were in and around those universities—doctors Chancey Juday, T. H. Langlois, Carl Hubbs, Karl F. Lagler, Albert Hazzard, J. W. Leonard, and Robert Ball, along with John Van Oosten, and others.[4] Few of them were present before World War II. Their early students began to graduate in the late 1940s and early 1950s. I was on the front edge of this second or third generation because I had nearly completed my undergraduate education in bits and pieces before and during World War II.

This group of educators and scientists, along with their students, produced much of our early understanding of freshwater fisheries. The research they published ultimately provided the basis of modern freshwater fisheries management.

I was very fortunate that this emerging science grew up very logically in the northern Midwest because of the abundance and importance of lakes and streams. The Great Lakes and their shorelines were a part of the fishing culture and its activities; therefore, the demand for management and understanding developed largely in this area.

As fisheries biology emerged as a science, biologists came to understand that introducing "new" species was almost automatically undesirable and, therefore, that intentional introductions should never be attempted. Their conclusion was reinforced by the example of the rainbow smelt described earlier. In more recent years, the sea lamprey and the alewife are cited as examples of what happens when aquatic invasive species enter accidentally. It was probably these elements in our historical background that provided a firm, almost unchallengeable, professional attitude that any new species—emphasis on any—is or will be, predictably, bad, bad, bad.

An unbiased assessment at that time would have had to conclude that there was more scientific understanding here, as evidenced by leadership and publications, than anywhere else in North America. My perspective, and I saw evidence of it, is that there was also a smugness, a feeling that we knew it all or at least more than anyone else and, in this matrix, we didn't need new ideas. We learned and were functioning within a particular environment as far as fishery resources were concerned. We had an enormous abundance of water. Our streams were stable, with little annual or seasonal variation in flow. Our fish populations were mainly natural and essentially intact, as far as original native fish populations were concerned. These populations reflected the stability, the abundance, and the productivity of our naturally occurring freshwater environments. Yes, there were a few dams and their accompanying reservoirs on some of our streams, but even these reservoirs were quite stable in water level.

I think I never sat down at that time and thought about the superiority of our science in the northern Midwest, but with hindsight and the perspective that comes when you can look back and compare, I think that's a fairly good assessment and description. I think that the firm dogma against introducing new species that prevailed in Great Lakes states came from the history of most introductions of new species in the last quarter of the nineteenth century. Unfortunately, the hatchery-oriented Fish Division supervisors in the Michigan Department of Conservation, as of the early 1960s, had not employed a scientific approach.

Michigan
Matters

Meanwhile in Michigan

n the twenty years following World War II, America experienced a period of prosperity, growth, and rapid change. Greater family incomes, combined with more leisure time, quickly led to an increased demand for recreational opportunities, including fishing, in those years. This demand was never precisely delineated, but it found expression in several ways and was a **critical factor**.

Multiple dynamics were at work. The inland lakes of Michigan were becoming dominated by boats with large motors, often pulling water skiers and sometimes endangering, but always irritating and disrupting, fishermen seeking quiet and solitude in addition to their usual fish. Similar competition was occurring on Michigan streams, as paddling canoes supplied by commercial entrepreneurs became superabundant and extremely annoying to anglers making quiet approaches and casting into the homes of the elusive brown and brook trout. These crowded conditions and the frustration that they brought resulted in fishermen traveling to lakes and streams in the wilderness areas of Canada, and in the wintertime to pursue saltwater fish in the states of our Gulf Coast.

During the twelve years that we had spent in Colorado, Michigan state parks had become overcrowded, and they were often inadequately maintained. Hunters clamored for doe permits to expand their hunting options, and the situation of the Detroit River, where thousands of ducks and geese had died from exposure to floating oil, contributed to a general feeling of discontent. Soon, the Michigan Conservation Department came to be viewed as stagnant and resistant to change, and there was a great deal of pressure on it to respond more quickly to the new economic conditions. But the mood of the nation was buoyant and optimistic, and most people were generally open to new ideas.

We in Colorado knew nothing about these developments in Michigan, but they were about to change our lives again and forever. A very significant one was the special 1961 convention led by John Hannah, then president of Michigan State University. Delegates produced a new state constitution that called for major changes in the organization of state government. Specifically, it stipulated that state agencies, then numbering over 110, be reorganized into no more than twenty departments.[1]

This mood, this sense of need for change, was visibly expressed in the gubernatorial election of 1962. During his campaign, George Romney promised that, if elected, he would "investigate the Conservation Department." He was, and he did. In the newly elected governor, Michigan had a leader dedicated to change—sweeping change, which became a **critical factor** for us.

Reviewing the Conservation Department

Romney appointed commissions over agencies, including one over the Conservation Department, which would promote change. He kept another campaign promise and that was to have a "blue ribbon committee" critically examine the Michigan Conservation Department and make recommendations. The department had in many ways become a victim of the nation's newfound abundance.

The Conservation Commission immediately called together a small group of people who had demonstrated dedication to the principles of scientific management of public resources. They appointed outstanding individuals in the field of conservation, including Robert C. McLaughlin, August Scholle, Ben East, George A. Griffith, and Judge Louis McGregor. In due time, the committee submitted a report that was extremely critical of the Department of Conservation and its leadership.[2] This would become a **critical factor** in our new approach to fisheries management.

The report criticized the Fish Division's emphasis on hatcheries and the qualifications of its leaders. It pointed out the division's lack of clear policy and

Director Ralph A. MacMullan
MICHIGAN DEPARTMENT OF CONSERVATION/NATURAL RESOURCES

direction, its budget allocation, its research programs, and the lack of a functional relationship between divisional leadership in Lansing and programs in the field. Unsurprisingly, after the commission's critiques were released, the news media labeled the division a "disaster."

The basic report was well-publicized, and Governor Romney called for its findings to be implemented. As one of the first steps in this, Gerald Eddy—an able geologist who had been director for thirteen years—stepped down to become chief of the Geological Survey Division, and the Conservation Commission chose Dr. Ralph MacMullan as its new director.

MacMullan was quick to act. He made some organizational changes and turned his attention to the Fish Division. He was eager to apply new, scientific management concepts to the division's work and to shake up the dysfunctional relationships that had been plaguing it for many years. This meant that the dominance of the hatcheries section of the division needed to be addressed. Thus, he came to grips with the following elements of its history.[3]

In the early days of fishery management in Michigan, the focus was on developing hatcheries and stocking fish. Most of our early hatcheries produced small trout, although bass, bluegills, and walleyes were also raised and released. This section grew to dominate the Fish Division, and both state and federal agencies built many hatcheries. Nearly all these early rearing efforts produced and released fish no larger than "advanced fry," which were young and less than two inches long.[4]

By the 1950s, research had shown that small trout from the hatcheries released into streams were not surviving to "enter the catch," meaning that they died before growing to seven inches (the general minimum size limit at the time).[5] Therefore, very few were caught by anglers. Since the whole purpose of the hatcheries was to enhance the public's fishing prospects, and since the hatcheries were consuming 25–30 percent of the division's annual budget, they represented an increasingly irresponsible use of resources.

The older, established hatchery group had occupied most of the administrative and policy sections for many years, and this meant that most budget decisions were made based on the needs of fish hatcheries. However, this situation entailed several problems, not the least of which was that the primary task of the hatcheries—to stock trout in trout streams—was simply not working.[6] It had produced very little in the way of improved fish populations or fishing in Michigan.

Additionally, hatchery personnel had never been required to have advanced levels of education. They were often simply promoted up the line, which only served to separate them from the more scientific and educated departmental employees. Moreover, since many hatchery personnel had risen to high positions, they were effectively protected from reform, much to the chagrin of the biologists in the two other main sections of the division—the Lake and Stream Improvement Section (LSIS) and the Institute of Fisheries Research (IFR).

The Conservation Department and the University of Michigan had worked together to form the IFR, which was staffed by many well-qualified and highly trained biologists. They had a sufficient budget to conduct good research and to carry out experimental programs. The university's fisheries faculty included several prominent educators and researchers—Drs. Hubbs, Lagler, Hazzard, and Gerald Cooper. Also located on the UM campus was the Great Lakes Fishery Lab of the federal Bureau of Commercial Fisheries. John Van Oosten was its longtime director, and he was a powerful force in the development of our understanding of Great Lakes fisheries dynamics.[7]

Besides the hatchery and research sections, the other component was the LSIS. For several years it was led by a flamboyant but effective leader named Horace Clark, who had been a fighter pilot in World War I. Picking up on some of the stream work done by the Civilian Conservation Corps (CCC) of the 1930s, the LSIS did some really good fisheries management work. When I arrived, Wayne Tody was leading this section, along with Roger Wicklund. They put many different structures into streams to provide fish cover and to stabilize banks. They provided stream fencing where grazing livestock were causing erosion, and they were also able to promote contour plowing to reduce runoff. These programs, in my judgment, were very effective and, for the most part, well worth their cost.

The IFR's leadership frequently asked for the opportunity to evaluate the LSIS program's product in terms of additional fish production or improved quality of recreation. The LSIS staff was asked to develop criteria that would express, in measurable terms, the overall improvement in lake and stream habitat that resulted from their efforts. Horace Clark had strenuously resisted these bureaucratic requests, declaring one time, "Hell's fire! Anybody can look at our product and see that it's good—why waste time and money evaluating the stuff."

Now, with the new director beginning the process of revitalizing the division and selecting a new chief, all three groups sought to be in the new leadership. Adding to the dilemma was the fact that the department had always, up to that time, selected division chiefs from within. In fact, I can't recall a single exception.

The first thing MacMullan did to deal with this situation was to appoint Dr. James McFadden—an outstanding young researcher from the ranks of the IFR—as acting chief. He did this knowing that McFadden would leave in six months to begin a faculty position he had already accepted at the University of British Columbia. The director ordered a plan for the Fish Division's future, and McFadden proceeded to produce it on schedule, naturally with considerable help from other staff, including Wayne Tody and others.

McFadden's Report

The McFadden report addressed all the division's needs in a systematic way, including expanding its horizons to include the Great Lakes.[8] Its major, basic recommendations were:

1. The legislature should recodify all the fisheries statutes.
2. Maintenance stocking should be approved only when growth would at least compensate for mortality of the maintenance fish stocked.
3. No stocking should be done where natural reproduction was adequate.
4. Introduction of exotic fish to new waters was acceptable wherever it was biologically justified.
5. Cut the fish hatchery program and release part of the funds, perhaps 40 percent, to other management projects in the division.
6. Promote the continuation of fish habitat work and improvement of the state's lakes and streams, not only with structures, but with population control and any other measures deemed necessary.
7. Work toward developing a legal means to restrict the number of commercial fishing operations, called "limited entry."
8. Establish enforceable rules to restrict the commercial catch.
9. Establish additional new Great Lakes areas, closed to commercial fishing, to enhance sport fishing.
10. Cooperate with the Great Lakes Fishery Commission on sea lamprey control and coordination of research with the other states, the Province of Ontario, and the federal fisheries agencies of the United States and Canada.

This set of recommendations formed an important foundation for the work the division needed to do. The report was presented to, and approved by, the Conservation Commission in June 1964.[9] It served notice on Michigan that the state's fishery management approach would be changing. It would be research-based and oriented toward delivering the best quality fishing experience to the largest number of citizens for the greatest length of time.

Responding to this report, Director MacMullan quickly displayed his very considerable talent for leadership, demanding action that, among other things, resulted in retirements. The second thing that the director did was to take the unprecedented step of advertising nationwide for people to take the exam for the position of fisheries chief. He apparently was convinced that the selection of a leader from within the ranks of that division would perpetuate existing conflicts. My returning to Michigan to assume the post as chief of the Michigan Fish Division was a result of these decisions.

Returning to Michigan

When I heard about the posting for fisheries chief, it was almost too late to apply. I got a phone call from Dr. Peter Tack, chair of the Department of Fisheries and Wildlife at Michigan State. He was one of the four or five people conducting the oral examinations of the applicants. He asked if I knew about the opening and, if so, why I had not yet applied. I told him I had not heard of it, and he told me that the committee had not yet found any qualified applicants. They had extended the search once and were now nearing the second deadline. After several conversations with Helen, I applied by telegram.

Colorado Considerations

It was not obvious at the time, though, that a position in Michigan was the best option for me and our family. Our overall situation in Colorado was good. We owned our home and liked the neighborhood where we lived. Helen was working halftime at the university and enjoyed it. She was also active in PTA and served as a Cub Scout den mother for nearly ten years. Our three boys—Mark (fourteen), Carl (twelve) and Hugh (ten)—were happy with friends and school. I was chief of fisheries research, loved most aspects of my job, and had a new office/lab building to work in—only five minutes from home.

Furthermore, I had a multitude of important and interesting projects underway. My staff included several of my former students and other respected and productive professionals. We liked the Colorado climate, the mountains, the university, and the town of Fort Collins. We had several couples of close friends with whom we enjoyed potlucks and bridge games. The hunting was great, and the fishing was pretty good.

Another element of our position in Colorado was stability. This, I believe, was especially important to Helen. When she was growing up, her parents had moved frequently. She had attended three different high schools and had always been "the new kid on the block." Our twelve years in Colorado were the first real stability of place that she had ever known.

There were, however, a few negatives about our Colorado situation. Salaries for state and university employees were low by comparison with other western states, and they were substantially lower than Michigan's. I had started work, with a new PhD in my pocket from MSC, as an assistant professor on a twelve-month appointment at $4,250 a year. After working there for nearly twelve years, I was making $10,800. When raises were discussed, we were often told "the scenery in Colorado is great"—to which we responded, "yes, but we've bought all the scenery we can afford!"

To compensate for low pay, Colorado Parks and Wildlife Department built some fringe benefits into our compensation packages. They issued me both standard and dress uniforms—quite military in appearance—that I wore on certain occasions. The uniforms came complete with badges and a big Stetson hat! They also provided me with sleeping bags, camping gear, waders, hip boots, and a car.

Michigan's Appeal

On the other side of the ledger, Michigan was home to both of us. Since Helen and I were our parents' only children, the fact that they lived in Michigan and were getting older weighed heavily on us. My father had already had a cancer operation, and Helen's father had Parkinson's disease. The rest of my family—aunts, uncles, and cousins—were also mostly in Michigan. Additionally, Michigan State College was my school. Moreover, both Dr. Tack and Dr. Bob Ball—my major professor and mentor from Michigan State—had urged me to apply.

One huge factor in the ultimate decision to go back was the abundance of Michigan's lakes and streams, a fact I had always taken for granted when I was growing up. For native Michiganders, the Great Lakes are just "there." We don't necessarily see them as anything extraordinary. Even though I had talked about the tremendous fisheries potential of the Great Lakes with professors and fellow students while going to MSC in the late 1940s, it hadn't amounted to anything more than theory.

My view of the possibilities had seriously shifted during my twelve years in Colorado. Colorado was a beautiful state with a great deal of biological diversity; yet, from a fisheries standpoint, it did not have a lot going for it. It was terribly short of water, and natural lakes were small and few. Streams often went dry or diminished to a trickle from irrigation diversions. Reservoirs fluctuated ten to twenty feet in depth as water was stored and then taken out for farms and cities. With a few exceptions, fisheries management was very difficult, and would inevitably become worse as human populations grew.

Granby Reservoir was the largest body of water in Colorado at that time, with a surface area of at least six thousand acres when full. When I made the Great Lakes decision, I was moving many, many orders of magnitude: from experience on waters of about six thousand acres to a decision that would permanently impact the biology, the allocation, and the economics of a huge area of water—more than ninety-four thousand square miles.

When we lived in Colorado, our family traveled to Michigan every other year, and such trips always renewed my awareness of the abundance of fish management opportunities in the state. On one trip, we rode a ferry to cross Lake Michigan, and that took six hours. We were out of sight of land for at least four of those hours. Now, you have to live in a water-deprived area like Colorado to appreciate the impact of that experience.

Back in Fort Collins, I would describe the wide expanse of Lake Michigan in my classes. It was difficult for students to comprehend that Michigan was surrounded by four such lakes. So, when I was later contemplating whether to return to Michigan to take management control of the state's fisheries, I thought of a quote from John Hannah, then president of Michigan State College, who said: "If you want to raise something, go where it will grow." When he said it, Hannah had been referring to agricultural crops and lands. In my mind, though, it applied just as much to fish and water. The clincher was that Michigan was recognized as a national leader in fishery research and professionalism.

The starting salary for the job was $12,000, a modest increase over my Colorado position. However, my uncle Richard "Dick" Allen, who was administrative assistant to the director of the Michigan Highway Department at the time, told me he was sure that the Michigan Conservation Department could hire at the top of the pay category, which was $15,000. This would be about a 35 percent increase in pay over my Colorado salary. In Michigan, I could also expect larger annual raises than I had received in Colorado. Thus, we decided to go for it.

We were scheduled to go back to Michigan in the summer of 1964, so I arranged to take the exam for the fishery position while I was there. It consisted of a review of my credentials and an oral exam. I remember that Dr. Tack, Dr. Gerry Cooper (director of the IFR and one of my supervisors during the summers of graduate research), and Chuck Harris, then a deputy director of the Michigan Conservation Department, were on the selection committee, along with some others I can't recall. During the exam, I was already thinking about how I would expand the Fish Division's management of the Great Lakes, and that I would seriously consider introducing kokanee salmon to inland lakes if I got the position.

Following the interview with the selection committee, I had a brief conversation with Deputy Director Harris. He told me that the committee would contact me no matter what its decision was. My family and I were leaving for my parents' home in a day or two, so I gave him the phone number for the sheriff's office and residence. To give a sense of how things have changed since then, the phone number was 30.

After several days, he called and said he wanted to meet with me. We arranged to get together at a hotel in Cadillac. It was during this phone call that I said I was interested in the position only if the department was prepared to start my salary at the top of the grade. Helen and the boys traveled to Cadillac with me, and they went out for ice cream while Chuck and I talked. It was a good conversation, and before we parted ways the position as Michigan's chief of fisheries was mine—at $15,000 a year. I was to report for duty in early September, before the start of school.

Making the Move

The rest of the summer was a blur. It was already mid-July, and I had to break the news to my colleagues in Colorado. I also had the difficult task of telling my friends, neighbors, and research teams. Besides that, we had to sell the house, conduct yard sales, and load a moving van, but we took care of everything. Before we left, we were guests of honor at a going-away party with gifts, dancing, and a load of best wishes and goodbyes.

After twelve great years in Colorado, we were heading back to Michigan. We had grown to love the West, and it was not easy to leave, even if we were heading back to what would always be "home country." For our boys, though, Colorado was the only home they could remember. Hugh was a native Coloradan, having been born in Fort Collins. So, many years later, as each son reached maturity and independence, they all moved back to Colorado, where they reside today.

We arrived in East Lansing and stayed a few nights with Bob and Betty Ball. Dr. Gene Roelof found us a house in Haslett to rent, and the boys started school. Helen was busy with a new household. Old friends and members of the MSU Department of Fisheries and Wildlife made us feel most welcome.

Director's Orders

I began my new job in the Mason Building in downtown Lansing. That seemed tall at seven stories, and Lansing seemed like such a big city to us then. The first days were hectic. I met many members of the department. Then I had my first meeting with Director MacMullan. I had first come to know of him when I was a graduate student, probably in 1950 or 1951. He was working full-time while finishing his doctoral dissertation, while I was working on mine. He was four or five years older than I and had been a bomber pilot in the Army Air Force during World War II, with many missions over Europe to his credit. He was about 5' 8" tall, stocky, quick-moving, and sometimes loud and profane. Since we both, over the course of our graduate work, met regularly with Dr. Don Haines—a statistician in the MSU Zoology Department—we occasionally encountered one another and exchanged small talk. We didn't have any significant discussions, though, until I moved back to Michigan.

It was easy to tell that Ralph was a dynamic leader. He welcomed me warmly, and quickly made it clear that he was making a lot of changes. He was blunt. He told me the Fish Division was a dysfunctional wrangle of three factions. He told me that nothing new had come out of the division in many years, and he told me to look carefully at the McFadden report. He spoke highly of Wayne Tody and a few others who would be my colleagues and made it abundantly clear that he expected me to move the division forward. I'll never forget his closing remarks: "Do something!" Then he added, "And if you can, make it spectacular!"

Commercial Fishing

Responding to the director's "spectacular" charge required serious consideration of, and action about, several fishery management issues, not the least of which was the status of commercial fishing. Just as we had taken the presence of the Great Lakes for granted, we had also taken for granted that they were the realm of commercial fishermen, with very little offshore sport fishing. That's the way it had always been, and that's just the way it was.

Yet, the commercial fishing industry in the Great Lakes was in a state of collapse. Fish had been a low-cost dietary staple during the Great Depression, which encouraged continued employment in an industry producing a cheap and readily available protein food. World War II brought increased demand and higher prices, which made fishing profitable for all species that could be caught.[1]

However, the increasing use of more efficient nylon nets contributed to overharvesting and a severe depletion in the fish populations—especially lake trout and whitefish—that had normally sustained commercial fishing. Moreover, predation by the recently arrived parasite, the sea lamprey, was amplifying the already drastic effects of excessive exploitation.

These conditions were combining in such a way that they eventually led to the final demise of the old style of Great Lakes commercial fishing, placing all

Commercial fishing dominated Michigan's Great Lakes fishery for almost a hundred years.
GREAT LAKES FISHERY COMMISSION

the operations either temporarily out of business or suffering from serious loss of income.[2]

Arguments for Change

Beyond the fact that the commercial fishing industry in the Great Lakes had been in decline for quite some time, we had more philosophical arguments for the change in allocation that we wanted to make. First, all Great Lakes fish populations belong to the public—the citizens of the United States and Canada. In these two countries, policies and laws regarding their conservation are the prerogative of each state or province. Migratory animals are further regulated by federal policy and law. Historically, the fish populations of Michigan's portions of the Great Lakes had been managed by the state's Conservation Department, which had managed according to the needs of commercial fishing operations. Those practitioners were subject only to regulations intended to prevent overfishing and depletion. However, overfishing had become chronic, and most fish stocks became depleted on a semi-regular basis.

An even deeper fundamental problem with this system was that commercial fishermen were harvesting and selling fish that belonged to the public, and the public was never really compensated for the sale of its fish. Some of the commercial operations, based on the privilege of harvesting a public resource, were quite large and included processing and shipping facilities on shore. Yet the only requirements they had to fulfill in the state's eyes were the purchase of a $25 commercial license and basic compliance with the existing laws.[3]

In effect, these commercial fisheries were practicing "market hunting"—taking animals from wild populations and selling them. In the early days of Michigan's non-native settlement, market hunting was fully appropriate and generally accepted. Gradually, however, as human populations grew, wild populations of animals declined, and almost all market hunting ceased.

At a certain time in Michigan's history, the allocation of Great Lakes fish to commercial harvesting might have fit my definition of conservation. In other words, it might have produced the greatest good for the greatest number for the longest time. However, it was now clear that to allocate the fish stocks of the Great Lakes to sport fishing would produce benefits that were several orders of magnitude greater, affecting hundreds of thousands of families, and we knew that the allocation primarily to commercial harvest had to end.

The allocation of assets from a public resource is probably the most difficult aspect of fisheries management anywhere. Allocation in any form is difficult, but reallocation—to take the assets and opportunity of a public resource from one user, in this case a commercial fishery, and to assign it to another, in this case a recreationally oriented fishery—is especially difficult and disruptive of established patterns. However, it is an important part of the Great Lakes salmon story.

By 1964, when we set about the difficult task of reallocating the fish populations of the Great Lakes from commercial fishing to sport fishing, perhaps a thousand families in Michigan depended on commercial fishing for a major portion of their income. Many others were taking advantage of the easy regulations and low fees to harvest fish indiscriminately, and thus to exploit a public resource for their solitary gain. The basic decision we made at that point was that the allocation of most fish populations to extensive commercial exploitation no longer made sense. This was of crucial importance, a **critical factor**, in creating a world-class sport fishery.

The idea of ending, or severely limiting, commercial harvest from the Great Lakes was not a great departure from what was going on in the rest of the country.

Decades earlier, most states had ended, or nearly so, commercial harvest from inland lakes and streams. The only difference that the Great Lakes posed was that they were much larger bodies of water and involved many more people. Of course, some might say that the larger size of the Great Lakes made them candidates for allocation to both commercial and sport fishing. In principle, nothing is wrong with that idea. However, we argued that even in the Great Lakes, equal coexistence of commercial and sport harvest was nearly impossible to carry out in practice.

Then came the announcement from the Michigan Fish Division, and especially from me as its leader, that there would be a reduction in the numbers of species available for commercial harvest and severe restrictions on where and how they could fish. The simple fact is that, once sport fishing was recognized as the best allocation of resources, it was necessary to restrict commercial methods of fishing so that they would have little or no impact on sport fishing. In conservation circles, this practice is known as managing for the "key value."[4] Recreational fishing would be the key value, and commercial fishing efforts could exist only where there was no or little negative impact on recreational fishing and the development of the largest and most valuable freshwater recreational fishery in the world.

We in the Fish Division were always depicted as anti–commercial fishing and were charged with wanting to eliminate it entirely, but that was never accurate. Our goals were both more complicated and more difficult to achieve. We envisioned that a financially viable commercial fishery could exist, taking species of little interest to the recreational angler—specifically lake whitefish, chubs, and in some instances, perch and walleye. However, to carry out this restriction, it was necessary to limit commercial fishing gear to that which would not take and kill significant numbers of those species allocated to sport fishing.

In practice, this meant that we had to eliminate the use of gill nets almost entirely and allow fishing only with trap nets. The difference between these two types of gear is that gill nets are not selective for species. Furthermore, fish caught in gill nets are usually dead or seriously injured when the net is lifted, making it impractical to release species that are no longer available to the commercial fisherman. However, when a trap net is lifted, a great percentage of the fish are alive and in good condition. In other words, trap nets are a selective gear. Commercial fisheries can retain those species allocated to them and release all the others.

Game Changer

A relatively small percentage of Michigan fishermen were already employing these forms of fishing gear and would be far less impacted by these restrictions. However, a large percentage of Michigan commercial operations were fishing with gill nets. For a commercial operation to switch from gill nets to trap nets was expensive. It is a very simple task, requiring little or no experience, to effectively set and lift gill nets. Further, it can often be done from a relatively small boat. Larger boats were used to set extensive arrays of gill nets, and these boats were generally enclosed with little open deck space. In other words, for a fisherman to switch the type of gear employed would often require a substantial cost for gear, larger boats, and the development of additional skills. Bottom line: it would work for some but not for all.

In addition to protecting the key value of sport angling, our restrictions had to provide rules and regulations that would leave the commercial operations the opportunity to fish profitably without overharvesting and depleting the fish stocks. Adequate laws and enforcement are easy enough to describe, but are very difficult to achieve in application.

The most important law that we needed to apply was "limited entry." Under such laws, regulating agencies have the power to limit the number of commercial fishing enterprises to license. At the time, most fishery managers understood the necessity of limited entry, but had achieved little limitation in practice. As far as we knew, the only successful application of these laws up to that time had occurred in the halibut fishery of British Columbia. There, the recovery of a badly depleted stock of halibut, with the help of catch quotas, set a precedent that has been repeatedly successful in numerous other commercial fisheries.[5]

The Fish Division began by establishing standards for defining a full-time, or bona fide, commercial fishing operation. We wanted to establish a basis for denying licenses to all others with the hope of drastically cutting the number of "commercial" fishermen on the lakes. The price of a commercial license was then raised to several hundred dollars, and commercial fishing operations were required to declare a home port. They were henceforth restricted to fishing within fifty miles of their home port. This would prevent fishermen from "swarming" to an area where an abundance of a desired species had been detected. In addition, most gill nets were no longer permitted. However, some small mesh gill nets could be set for chubs in very deep water (over two hundred feet), provided monitoring officers could confirm that forbidden species were not being taken, or taken only rarely.

Commercial fishing interests vigorously opposed all these restrictions in court. They generally contended that the Department of Conservation lacked the authority to impose them. The various legal battles carried on long after I left my position as chief of fisheries. They dragged on because the commercial fishermen who had lost in the lower courts (as they invariably did) would appeal and use every legal stratagem at their disposal to create more delays in implementation. In fact, some of the suits were not ultimately defeated in the Michigan Supreme Court until after I became DNR director in January 1975. This meant that, during those years, a certain amount of the old-style commercial harvesting continued.

Resistance

However, resistance to the implementation and enforcement of these new regulations did not always happen in the courts. Some fishermen frequently resisted or surreptitiously avoided conservation officers' inspection of boats, nets, and catch. Some instances even involved violence. Once, a conservation officer was thrown overboard; at another time, patrol boats were the target of rifle fire from shore. The Garden Peninsula, projecting south from the Upper Peninsula into Lake Michigan, and the adjacent waters of Bay de Noc were particularly difficult areas.

Conservation officers armed with search warrants frequently raided commercial fishing boats and fish processing facilities. After one raid, at least one large commercial fishing operation was charged with and convicted of income-tax evasion. More often, fines were levied when violations of fishing regulations could be proven in court. However, soon it became clear that fines were not an effective deterrent. Therefore, we began to seek long-term suspensions of fishing licenses, or, in the case of repeated violations, permanent suspensions.

During one period (after I had left the division), suspicions grew that a source within the Natural Resources Department was warning commercial fishermen about impending searches. After some investigation, the source of the leak was found to be an individual very highly placed within the department's administration, although not within the Fisheries Division itself. Once discovered, the leak was ended by a forced early retirement.

After all the years of resistance, those commercial fishing enterprises that remained emerged as stronger, larger, and more profitable than they had been during the less-restricted era. For when unlimited numbers of commercial licenses were issued, fish stocks had been generally depleted. We called those fishing licenses "licenses to starve." With fewer commercial fishing operations, the fish were more abundant, and the catch rates were higher. This had the effect of lowering the overall cost of fishing while maintaining a strong market demand for the catch. In other words, our efforts brought stability to the smaller commercial fishing industry and greater prosperity to the fishermen themselves—even though we had to bring them there kicking and screaming all the way.

The Great Lakes Fishery Commission

he collapse of the Great Lakes commercial fisheries in the 1940s created such
a crisis that it finally brought the governments of Canada and the United
States together to make a serious joint effort to take control of the situation.
After more than a decade of negotiations, the result was the international body
known as the Great Lakes Fishery Commission (GLFC), created in 1955.[1]

The commission was established with dual provenance, and its headquarters
office was in the same building as the U.S. Bureau of Commercial Fisheries in Ann
Arbor, with the bureau becoming its action arm in the United States. It is funded
jointly by the two countries: 61 percent by the United States and 39 percent by
Canada. This division was developed because 61 percent of the surface area of
the Great Lakes is within the boundaries of the United States. Lake Michigan,
which lies totally within the U.S., creates the disproportionate split, whereas the
international boundaries that divided the other lakes relatively equally had been
in place since the end of the War of 1812. The directorship of the staff alternates
between the two countries.[2]

As originally conceived and proposed, the Great Lakes Fishery Commission
was to have management authority over all the waters of the Great Lakes. However,
another **critical factor** in the development of the Great Lakes sport fishery was the

The United States and Canada worked to solve fishery problems through the binational
Great Lakes Fishery Commission.
GREAT LAKES FISHERY COMMISSION

issue of balance between regional/national power and state jurisdiction. Under U.S.
law and by tradition, the management authority over populations of wild fishes
and game animals, with a few exceptions, had been awarded to the states, not to
federal government agencies such as the U.S. Fish and Wildlife Service. However,
while each state retained management authority, the federal agencies exercised
what I would term de facto management, principally because the U.S. Bureau of
Commercial Fisheries, eventually succeeded by the U.S. Fish and Wildlife Service,
conducted most of the monitoring and research that examined the status of Great
Lakes fish stocks.

For the Great Lakes Fishery Commission to gain such legal management
authority, each and all of the eight Great Lakes states would have had to voluntarily
surrender their management rights and formally transfer them to the commission.
Their counterpart in Canada—the Province of Ontario (the only one riparian
to the Great Lakes)—would have had to do the same. All the Great Lakes states
and Ontario were formally requested to transfer management authority to the
commission, and each complied, save one—Ohio. Because Ohio refused, the GLFC

was mandated only to "lead cooperative management efforts of the various states and Ontario."

One of the reasons for Ohio's reluctance might have been that the fisheries of Lake Erie had been least affected by the sea lamprey. The most important species in Lake Erie were walleye and yellow perch. Whether it was for this reason or others, Ohio declined, and I have always been grateful to them for this.

If Ohio had complied with the request to transfer management authority, the management of the Great Lakes would probably have remained keyed towards restoring profitable commercial fisheries. At the time of the salmon introduction, the commission was completely dominated by people from the commercial fishing industry, and their action staff members—the U.S. Bureau of Commercial Fisheries at the time—were completely oriented towards the support of the Great Lakes commercial fishing industry.

If Ohio had not successfully resisted the transfer of management authority, we in Michigan's Fish Division would have been confined to our state's inland waterways. We would never have been able to manage the Michigan waters of the Great Lakes in any meaningful way. I would not have had the authority or the opportunity to propose and achieve the introduction of Pacific salmon with the goal of reallocating the fisheries of the Great Lakes from commercial fishing to sport fishing. Another **critical factor**, indeed. So, hail to Ohio!

Tribal Fishing Rights

Another significant element in the Great Lakes fishery at that time was the position of Indian tribe members. It is very difficult for me to write about the legal struggle that the State of Michigan undertook to gain control over the commercial fishing activities of those groups who also possessed treaty rights to the fishery.

Tribal fishermen from the Bay Mills Indian Community, who were denied commercial fishing licenses during the program to reduce the number of those licenses, filed suit against the State of Michigan in the late 1960s, claiming treaty rights to fish without restraint from state or federal authorities. This began a long series of court battles with bitter defeats for those of us in state government and our many supporters in the conservation and sport fishing communities. Continuing throughout most of my tenure as DNR director, we lost one court battle after another.

It was difficult for someone like me, who has always seen Great Lakes fish to be a public rather than private resource, to accept what happened. But I do understand the reasons for our defeats. It seemed that most of the "commercial fishing" done by tribal members generally involved a small open boat with a few hundred feet of gill net, and it generally occurred close to shore in weather that permitted the operation of vessels that size. Oftentimes, the extended family and other members of the band used the landed fish. Occasionally, trade or barter might have been involved. However, this type of fishing was closer to subsistence than it was to harvesting for sale in wider markets.

It became clear that the Fish Division made the key error—that is, restricting those who could fish and the type of gear that they could utilize. When individuals reporting catches of fish of less than $5,000 a year were denied commercial fishing licenses and when gill nets were banned, we essentially made the fishing efforts of most tribal members illegal. We attempted to rectify this error many times by providing exceptions that would have permitted the tribal bands to fish in their accustomed ways in waters adjacent to their reservations. Those actions triggered the lawsuits brought first by the Bay Mills band and subsequently by all the tribes affected by the treaties in question.

Going to Court

Albert "Big Abe" LeBlanc and his elder brother Arthur of the Bay Mills Indian Community led the first lawsuits.[1] Both were intelligent though not formally educated, and both were strongly devoted to the well-being and interests of their tribal members. Soon after they filed the initial suit, our problems escalated. The action taken by the tribal bands quickly attracted strong and talented legal support.

A group of attorneys headquartered in Boulder, Colorado, and an organization—Attorneys for the Legal Defense of Native American Rights, now called Native American Rights Fund—entered the case on the side of the tribes. Some of the largest charitable foundations in the country supported this organization.[2] The U.S. Bureau of Indian Affairs (BIA) also provided attorneys and legal counsel. I characterized their goal as attempting to redress all the wrongs suffered by the various tribal groups that have occurred throughout the European American settlement of the North American continent.

As a further complication, some commercial fishermen who found their fishing activities restricted by our new management regulations also had blood rights in the various tribal bands. Those who had capital, large boats, extensive fishing gear, and long experience joined their complaints to those of the tribal groups who were claiming the right to fish without restrictions, thereby circumventing our ability to manage them.[3]

Losing the Legal Battles

The court battles went on for years. On our side, we had the part-time services of one lawyer from the Michigan Attorney General's office, and several nongovernment groups, including the Michigan Steelhead and Salmon Fishermen's Association and Michigan United Conservation Clubs, supported us. Various individuals also provided financial support so that we eventually had several hundred thousand dollars to spend. Still, we would usually have only one or two attorneys at our table, while at least half a dozen attorneys would be at the other with support staff hovering nearby. We lost every time. Our final appeal was to the U.S. Court of Appeals in Cincinnati, where we lost for the last time, sealing our complete defeat.

The courts basically said that the original treaties between the United States government and each tribe were treaties between sovereign nations, and that our nation must honor the language of the treaties. They ruled that the interpretation of the treaty language had to mean what the tribal leaders who signed the treaties would have understood them to mean at the time of the signing. Furthermore, the courts ruled that the term "waters adjacent to the various reservations" included all the Great Lakes waters within the area covered by the various treaties. Finally, the treaties often granted tribal members the right to fish and hunt "until the area was needed for occupancy." In our case, the courts ruled that sport fishing did not constitute a need for occupancy of the water. We repeatedly attempted to have Congress act to limit these court decisions, arguing that good resource management required compliance by all participants, but these efforts got us nowhere.

Of course, I agree that we had a long record of mistreatment of Native Americans. I agree that we need and have ways and means to redress some of these past wrongs. I vehemently disagree, however, with the courts' actions in these cases. The courts decreed that, in effect, eligible tribal members have superior rights to all the other citizens of this country: the right to fish in any manner, in any place, and at any time of their choosing.

Such rulings had important effects on Michigan's efforts to manage Great Lakes fisheries. The conservation ethic that we were trying to implement was simply an insignificant factor to those who were making these decisions. However, their decisions are the law of our nation, and we are a nation of laws. I totally disagree with the decisions and with the laws that were put in place. Yet, as a citizen I must heed them.

Bay Mills Indian Community conservation officer visits commercial fishing vessel.
B. SCHOFIELD—GREAT LAKES FISHERY COMMISSION

Today, so many years later, tribal fishermen continue to exercise their right to fish in the places and in the manner of their choosing. Their potential impingement on the best possible management of Great Lakes fishing is significant. However, it isn't as serious as it might be, mainly because they have exercised restraint and developed regulations.

The tribes have hired several competent biologists, and some have cooperated with state agencies and participated in various interstate cooperative management strategies, including the Great Lakes Fishery Commission's lake committees. Presently, trap nets are being increasingly used by tribes close to Manistee, apparently financed by money from their casino. Also, tribal cooperation in sound fisheries management has evolved through the consent decrees developed through the courts and implemented over the past thirty years.[4]

Fishes of the
Great Lakes

Lake Trout

O f course, we had other fish to consider in our management ideas when we decided to take advantage of the coho offer. The Great Lakes are home to dozens of established fish species, at least three of which—the lake trout, the sea lamprey, and the alewife—were significant at the time we seized the opportunity to do something even more spectacular than we had thought possible. The native lake trout deserved primary consideration.

From geologic time, since the glaciers retreated, the lake trout has been an important, highly desired native Great Lakes species. It has been a top predator in the biological systems of food and fishes in Lakes Superior, Michigan, and Huron.[1] Several aspects of the lake trout's status are important. Its biology is complex. The U.S. Bureau of Commercial Fisheries intended it to be the solution to restoring the vitality of fish stocks. The financial well-being of the commercial fishery was also critical.

Because the lake trout was the lakes' top predator when we in the Michigan Fish Division chose to introduce salmon, we (especially I as the division spokesperson) have sometimes been labeled anti–lake trout. I do not consider myself anti–lake trout and, of course, I will write about this without bias. At least I will try.

We were aware of major factors about lake trout—some biological, some management, some political, some perhaps sociological—that we might consider.

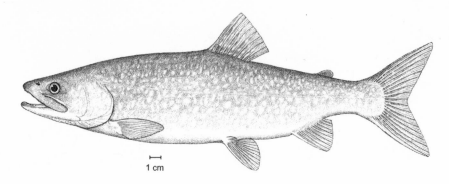

Lake Trout

Biological

Under present conditions, many biological factors negatively impact the lake trout in all the Great Lakes but Lake Superior. Lake trout populations of the other lakes are at the southern fringe of their native range. All animals existing near the limits of their natural range are vulnerable to climatic and other kinds of changes. Within the Great Lakes Basin, only the lake trout populations of Lake Superior are securely within the limits of their natural range. I know of no naturally occurring populations south of the lower four Great Lakes.[2]

Furthermore, fisheries biologists generally believe that portions of the lake trout populations were eliminated before adequate records were kept. Separate "races" or subspecies of lake trout in Lakes Michigan, Huron, and possibly Lakes Erie and Ontario likely disappeared in the early years of European American settlement.

The basis for these thoughts is as follows. Many lake trout populations that are scattered throughout the more northern parts of their native range in Canada are stream spawners. Some lake trout apparently had spawning runs up the streams flowing into Lakes Michigan and Huron, where they were stymied by the presence of dams. These races of lake trout, if in fact they did exist, could have been lost when early netting capability was developed. Nets placed across river mouths would probably have been enough to remove them. This loss would be one of the early events that weakened the lake trout populations in those areas.

One basic precept of fisheries management is to protect fish until they reach maturity and, as a year class, have an opportunity to spawn at least once. The lake trout is a very slow-growing species. It has been proven many times over that few, if any, young hatchery-reared lake trout survive unless they are raised to a size of at least six or seven inches before they are released into the lakes.

In Lake Michigan, it takes about seven years for a female lake trout to reach maturity, probably longer in Lake Superior. After about the fourth year, the fish probably exceeds twenty inches, large enough to be caught by nets. Even if it escapes that fate, it will still have been vulnerable to commercial and sportfishing gear for at least two years before it reaches reproductive age. Sea lamprey attacks also begin well before this trout reaches the initial spawning age.

The modern Great Lakes lake trout also has a basic vulnerability in its spawning. It naturally spawns on a variety of reefs deep within the lakes. When the trout hatch, emerging from protected locations in those refuges, they are very small and nearly helpless, still with an extended sac full of nutrients known as the yolk sac. This pattern of spawning was perhaps sufficiently efficient in early times, but is far less so today.

The young trout are highly vulnerable to any predator—crayfish, sculpins, and others. With the introduction of the smelt and later the alewife, not to mention more recent invasive species, the intensity of predation on the sac fry and the small fry stages increased dramatically. In recent years, the invasive quagga mussel has covered spawning reefs, perhaps a more indirect negative influence, but certainly another factor that reduces the survivability of young lake trout.

Another reason was the species' benthic location—its tendency to exist primarily close to the bottom and prey on the food species there. That tends to make it a less effective predator in the pelagic zone. However, such a designation in biology is not precise, and we learned later that lake trout, like most fish, are opportunistic and are occasionally found high in the water column or pelagic zone and feeding on the available alewife.

The negative effects of these factors are very clear from the multitude of attempts to restock the lakes with lake trout from hatcheries. So how do you protect the species until it has an opportunity to spawn?

Management

In addition to the biological handicaps, we also had management concerns about the lake trout. It is difficult because it is very challenging to get an early estimate of spawning success in any one year. Because the lake trout spawns on reefs in deep water, the resulting fry and/or fingerlings were not detectable by sampling gear for some time. Management is dependent upon its ability to predict future abundance, and again, that is very challenging in the case of the lake trout. Index netting of four- to seven-year-old fish might have provided an indication of upcoming-year classes, but we didn't do that then. Also, according to history, tradition, and management rules, the lake trout was firmly established as a commercial species; it would have been more difficult to reassign this principal part of a management scheme where sport fishing would be the key value.

All the above is less true of the lake trout populations of Lake Superior, though at that age they are equally difficult to assess. In this natural lake trout habitat, the species was less seriously affected by the smelt and the more recent alewife. In my opinion, self-sustaining lake trout populations in Lake Superior can be reestablished or extended and sustained by good management.

Restocking

As it gradually became apparent that it would be possible to greatly reduce predation by sea lamprey, it was clear that the door was also open to restocking lake trout. The U.S. Bureau of Commercial Fisheries looked for a suitable site or sites to build hatcheries to rear lake trout for stocking. One of these locations was on my beloved Jordan River—the central point of my early brook trout fishing, the first "wild and scenic river" to be established in Michigan, and a river and valley that were almost completely in public ownership. Long before my return from Colorado, the State of Michigan had donated the chosen site in the Jordan Valley to the federal government.[3]

In the summer of 1965, I appeared on the speaker's platform at the opening ceremonies of the new federal hatchery constructed on that site. On that day, I went to the hatchery for the first time. When I arrived, lo and behold, it was precisely on the location of the old stone house, one of the principal landmarks in my field of memories. I listened to the welcoming speeches, gave my own brief remarks, and drove a sentimental journey along the river. Technically, the site was fine, and the water supply was adequate. That hatchery is still functioning well today in its primary role of restoring lake trout to the Great Lakes. Others have since joined it, but it was the first and probably remains the most significant.

I still wish it weren't there. It represents the only significant intrusion in an otherwise natural watershed. Not many such watersheds still exist in Michigan's Lower Peninsula.

Early in the program to restore the lake trout, the government found it difficult to locate a native brood stock. By that I mean a source of sexually mature, bona fide lake trout that were guaranteed to be native to the Great Lakes. Since its very early years, the Michigan Department of Conservation had a facility near Lake Superior known as the Marquette Fish Hatchery. During these events, the hatchery superintendent was Russ Robertson. In previous years, the department had maintained a small lake-trout stocking program, and the trout they used were the product of the brood stock held at the Marquette hatchery.[4]

However, at some point the department had decided to stop stocking lake trout, probably because of the sea lamprey. As I understand it, Russ Robertson was ordered to dispose of his brood stock; they would no longer be needed. They occupied space and they cost money to feed. Russ decided to disobey the order. He moved the brood stock to a remote portion of the hatchery and probably reduced their number. And so it was that, when a brood stock was needed for the new hatchery on the Jordan River, Robertson's disobedience and foresight was the beginning of a Great Lakes brood stock.

Political

Now we come to the focal point of change, and the lake trout was very much a part of it. We in the Michigan Fish Division had chosen and announced our intent to develop a program and a direction contrary to all that had existed throughout history up to that point. We announced that we were going to build the world's greatest freshwater sport fishery and that, to do that, we were going to begin introducing Pacific salmon. Very basic to all of this was that, for the first time in recent history, the Michigan Conservation Department would manage Great Lakes waters within Michigan. Furthermore, while we may not have announced it explicitly, it soon became apparent that we intended to regulate the existing commercial fishery much more strictly.

The lake trout had long existed as a species central to commercial fishing. It was well established in law as a commercially allocated species. Tradition alone placed it in the minds of people as a species to be harvested for the market.

Sociological (Angler Interest)

The last factor that I want to mention will be familiar to most anyone who has been fishing on the Great Lakes. Lake trout are not highly regarded as a sport fish. Simply put, they do not put up much of a fight. I'm sure that if you ask anglers who fish the Great Lakes to list the available trout and salmon species in order of their preference, lake trout will rank low on the list. Keep in mind also that, at the time we made our decisions, what sport fishery had existed for lake trout relied upon a very long metal line with a heavy weight trolled on the bottom.

Fundamental to our plan to change the allocation of the Great Lakes fishery resource from commercial fishing to sport fishing was that we needed fish that were exciting enough to catch so that people would take the time and spend the money required to participate in fishing on the Great Lakes. The lake trout simply would not have been sufficiently exciting to support a sport fishery.

Perhaps it's time to say again that in choices relating to the proper management of a public resource—in this case a fishery resource—the manager should strive to achieve the greatest good for the greatest number of people for the longest time. We argued then, and subsequent events have proved us right, that sport fishing produced more personal, individual angling benefits and more economic and other benefits than any allocation to commercial harvest that we could have made. In another sense, the lake trout would have been cast in a role contrary to our intent.

My next statement is probably quite predictable. Our opposition would say repeatedly, "But the lake trout is our native. You are introducing an exotic." We had to respond to that argument repeatedly, but that's another part of the story and I will come to that.

Sea Lamprey

The sea lamprey, because of its parasitic attacks on lake trout and some other species, has been cast in the role of the archvillain, as the principal cause of the total collapse of commercial fishing in the upper three Great Lakes—Superior, Huron, and Michigan—in the years following World War II. Professional biologists have a substantial difference of opinion about these developments. Some attribute the fisheries crisis totally to the explosion of sea lamprey abundance that occurred in those years, while others point to the quickly expanding commercial fishing effort that occurred at the same time and with the use of much more efficient nylon nets. However, most would agree that the arrival of the sea lamprey—this deadly, invasive parasite—into the upper three lakes was a monumental moment in their biological history.[1]

In my mind, both the expanded and improved efficiency of the fishing gear and the sea lamprey predation share the blame, and I'll not try to assign percentages to either one. Regardless of how much of the blame should fall on the sea lamprey, I think it is important to have some understanding of this species, what a serious problem it represented then and remains today. Let's look at some background.

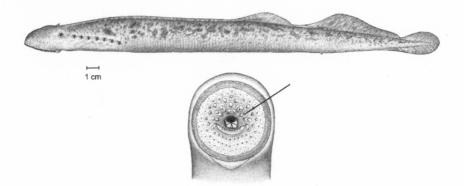

Sea Lamprey

Lamprey Characteristics

First, the sea lamprey, although a fish, lacks true jaws. Instead, it has a sucking disc equipped with an array of rasping teeth around its mouth.[2] The mature sea lamprey usually, but not always, swims in search of prey near the bottom of the lakes. Its ideal target is a fish larger than twelve or fourteen inches, preferably with thin, small scales that provide the least barrier to its rasping teeth. With its sucking disk, the sea lamprey attaches firmly to the side of a fish and can usually remain attached despite any thrashing movements that the fish might make to rid itself of the lamprey. Quite rapidly, its teeth open a wound about the size of a quarter, and then the lamprey proceeds to suck blood and other body fluids from its host. It may remain attached for a long time and most often, but not always, kills the fish.

From the perspective of the sea lamprey, the lake trout makes an ideal host. Its scales are small, thin, and provide the least protection of all the other large fish species in the upper three Great Lakes, with the possible exception of the burbot. The other trouts and several species of salmon have scales slightly more resistant to the attacks. However, these differences in vulnerability are not terribly important. The sea lamprey frequently attacks all of them. It also preys on lake whitefish and any of the other larger fishes and successfully attacks many other species, though less frequently.

Several species of lamprey live in the oceans of the world, and five—three of them parasitic—inhabit Michigan waters of the Great Lakes. The sea lamprey is widely distributed and can reach three feet in length. It had once been an anadromous species that lived in salt water but returned to fresh water to spawn. Like other species in this category, the sea lamprey entered North American streams that were tributaries to the North Atlantic.

Lake Ontario

Although its history is incomplete, I think we can assume that the sea lamprey was always in the Hudson River system and probably came up the St. Lawrence and occupied Lake Ontario, spawning in its tributaries.[3] However, it was not, apparently, a serious threat to the splendid Atlantic salmon population there until early European settlers exterminated that species with harvesting and habitat modification.[4] Like other species, the sea lamprey was blocked from the upper portions of the Great Lakes system by Niagara Falls.

I have one serious question about the assumption that sea lamprey was the primary detriment. Lake Ontario has very similar characteristics of temperature, clarity, and clear-running tributaries and so provided the basis for a substantial sea lamprey population. However, history documents that prior to European American settlement of the watersheds of Lake Ontario, the highly regarded Atlantic salmon was repeatedly described as very abundant.[5] The question becomes: how could the Atlantic salmon—a very good host target of the sea lamprey—become so abundant in the face of what was probably a thriving sea lamprey population?

A similar question has been raised in Lake Champlain or the Finger Lakes of New York because a lake trout population coexisted with sea lampreys and successfully maintained its abundance. How did this happen? These thoughts are the basis for my doubting that the collapse of the lake trout and the disruption of the food web of the upper three Great Lakes were totally caused by the invasion and population explosion of the sea lamprey.

Around 1830, the Welland Canal near Niagara Falls was completed, providing passage for ships beyond the falls into the other four Great Lakes. About the same time, the Erie Canal connected the Hudson River system to Lakes Erie and Ontario. In addition to providing passage for ships and early immigrants (some of my ancestors came by this route to Michigan), these canals provided access for sea lampreys to the upper lakes.[6]

The sea lamprey found its way into the waters of the other four lakes, but, for some reason, did not become abundant in the upper three lakes for nearly one hundred years. The most likely explanation is that Lake Erie and its tributaries failed to provide the clear waters, lightly scaled prey species, and cool, gravelly spawning sites that the sea lamprey needs to thrive.

Upstream

But beginning in the 1930s, sea lampreys began to appear in Lake Huron, then in Lakes Michigan and Superior. Their waters are well within the temperature range of the sea lamprey's original home. The sea lamprey requires a stream environment for spawning, and a stream suitable for resident trout species is near ideal. This is important, because more than one thousand streams flowing into the three lakes have been identified as suitable for lamprey spawning. The physical character of these lakes, with their clear waters (sea lamprey feed by sight), hundreds of streams with suitable gravel bottoms, and abundant food supplies of lake trout and white-fish, provided everything necessary for an explosion in sea lamprey populations.

The following personal experience may illustrate some points. In the summer of 1942, while serving as a cabin leader for the YMCA camp on Torch Lake in Antrim County, Michigan, another cabin leader and I took the eighteen or twenty 12- to 13-year-old boys by ferry boat to Beaver Island in Lake Michigan. One day during our visit, I obtained permission from a commercial fisherman for some of us to ride on his boat while he lifted gill nets. At the time, I didn't appreciate the significance of some things that I observed that day, but I would recall them many times during my professional work in Michigan.

We left St. James Harbor and proceeded to an area south and west of High Island. We spent most of the day running along what may have been miles of gill net. The nets weren't lifted completely, just raised, fish removed, and reset. When he finished raising the nets, the captain was obviously disappointed. He had about three boxes of whitefish and fewer than ten lake trout. One of the lake trout had an open wound caused by a sea lamprey attack, and the boat captain pointed this out to us. In looking back at these observations, he probably didn't make expenses that day.

What significant points did we learn? That experience probably was an example of the decline in Great Lakes fisheries that had already begun by 1942. Secondly, lake trout were already quite scarce and were already suffering at least some predation by sea lamprey.

Growing evidence pointed to a serious sea-lamprey problem in the upper lakes prior to World War II, but the war disrupted fishing and the collection of data about the fisheries. Following the war, though, its widespread, ravaging attacks quickly became apparent as the populations of lake trout and whitefish soon collapsed in Lakes Huron and Michigan. Severe declines in Lake Superior came later.

Controlling Substance

What could be done to control this voracious parasite?

The efforts, under the direction of Dr. Vernon Applegate of the U.S. Bureau of Commercial Fisheries, proceeded in two principal directions. One was to determine which streams tributary to the Great Lakes were sea-lamprey spawning sites, and to build structures that would block these runs. This was a localized, focused program of examining the life history, the factors necessary for spawning success, the length of time that the young (larval) sea lamprey (known as ammocetes) remained in the stream, and all other factors of the life history that might reveal vulnerability to control programs.

It was clear to every informed biologist that if the lamprey were to be controlled, it would be at those spawning sites and streams. Thus, the riparian states and Ontario were required to survey all their streams for evidence of sea lamprey spawning. This was done in 1947 and 1948 by a staff of biologists under Dr. Applegate's leadership. As I described earlier, I was personally involved in these survey efforts, working with Walt Crowe in the summer of 1947. The results showed that the sea lamprey spawned in all available stream areas that possessed the characteristics of trout habitats.

I met Dr. Applegate when he came to Michigan State's fisheries lab in the basement of Morrill Hall to recruit summer help for the sea lamprey project. He was a small man, slender, red-haired, and always very intense in everything that he did. Under Dr. Applegate's leadership, experimental blocking weirs were designed, modified, improved, and placed in selected streams for testing. He signed up Al Brower and Bill Bain, and they operated the experimental lamprey barrier on the Carp River west of Mackinaw City during the summer of 1947.

Dr. Applegate deserves most of the credit for achieving control of the sea lamprey, leading all the efforts of both the United States and Canada to cooperate in the development of an approach with the capability to control its populations.

He and his team did an outstanding job of determining the sea lamprey's life history, which allowed them to make better assessments about its possible vulnerabilities.

After completing surveys of lamprey spawning and other initial surveys that helped to define the problem, Applegate headed the Hammond Bay laboratory of the Bureau of Commercial Fisheries. This lab searched for a chemical that would kill the larvae of the sea lamprey without killing other fish. It was tedious and exacting work, but, after examining over six thousand chemicals, he and his staff were finally successful. As a result of his work, TFM (trifloromethyl-4-nitrophenol) and related chemical lampricides became the primary factors in achieving the high level of lamprey control that we have today.

When the Great Lakes Fishery Commission was formed in the mid-1950s, it was charged with ridding the lakes of this scourge and, soon after its establishment as a functioning agency, began large-scale control efforts. Subsequently, research developed improved techniques for applying lampricides to kill the sea lamprey while it is still in the ammocete stage in the streams. Today male sea lamprey are being captured, sterilized, and released to suppress the viability of sea lamprey spawning. Ultimately, a pheromone that attracts sea lampreys was identified and synthesized.

Some Success

Discovering the means of controlling the sea lamprey took many years of hard work. Efforts were designed to begin on Lake Superior, and by 1959, convincing evidence had accumulated that they had achieved effective control there. This was important because the restoration of the Great Lakes in any form would have been impossible apart from sea lamprey control. By the time I returned to Michigan in the fall of 1964, it was clear that Dr. Applegate's efforts and those of the people under his direction were going to achieve levels of success necessary for optimism about adequate survival of lake trout.[7]

This brought together yet another fortuitous circumstance, for if we hadn't had confidence that sea lamprey control was possible, we would never have been able to go forward in 1964 with acquiring the first coho eggs. I believe that if we had not had a reasonable expectation that sea lamprey control was feasible, we would have backed off, because we probably would have concluded that creating a population of salmon would have simply been adding more food for the sea lamprey. This is hindsight to be sure, but I do know that efforts to restock lake trout in Lake Superior were not begun until it was assured that the sea lamprey population there was being controlled. Had the control of sea lamprey not been achieved and recognized, any effort to introduce Pacific salmon would no doubt have been viewed as very unlikely to succeed and would probably have been denied at some point. It was definitely a **critical factor** for us.

Both the United States and Canada continue to conduct large programs to suppress sea lamprey in the upper three lakes. Substantial success has been achieved—not to eradicate, but to substantially reduce the numbers of sea lamprey and to reduce their predation capacity. Clearly these efforts in Lake Superior are sufficiently successful to permit a strong resurgence of lake trout populations there.

Growing evidence in Lake Huron suggests that lake trout are, for the first time in decades, spawning successfully and contributing to establishing and maintaining a self-sustaining lake trout population. These successes permit us to postulate that similar success would be possible in Lake Michigan and that all of this would come to pass.

However, in closing I must raise this question. If we accept as a fact, and I think we must, that there is no possibility of fully eliminating sea lamprey from the upper lakes, then how long will our two countries continue the very substantial expenditures of money to keep the sea lamprey suppressed to acceptable numbers? I believe that a day will come when a decision will be made to cease.

Alewife

With the lake trout seriously depleted—or perhaps it should be termed "commercially extinct"—the whole food web of these lakes was without sufficient predators to control the prey species, particularly the non-native, invasive alewife. The alewife is important to this story because we would count on it as a food source for the Pacific salmon as it was unbelievably abundant.

Origin

The original range of the alewife is the North Atlantic Ocean, principally from the area of Cape Hatteras, North Carolina, northward to, perhaps, Newfoundland.[1] The anadromous form of alewife, by the then prevalent definition, is a species that lives in salt water but enters freshwater rivers to spawn. A "landlocked" variety has developed the ability to live its life cycle in fresh water.[2] The alewife is extremely abundant at times, is principally a forage species (though possibly also predaceous on young fish of other species), and when harvested, its market value has been as a source of fish meal and fish oil.

It is a small fish, rarely reaching eight inches in length, and it also is relatively

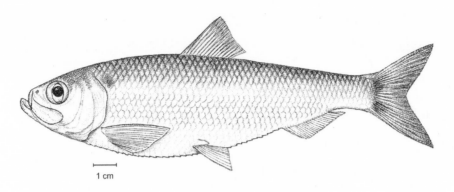

Alewife

AN *ATLAS OF MICHIGAN FISHES*, USED WITH PERMISSION FROM THE UNIVERSITY OF MICHIGAN MUSEUM OF ZOOLOGY

short-lived. Its life expectancy is no more than nine years.[3] You may remember from history lessons that the early Puritans' Indian contact Squanto taught them to plant and raise corn by burying fish next to the young plants—a practice he likely learned from a visit to Europe. It's possible that that species was the alewife.[4]

Turning now to the presence of the alewife in the Great Lakes, very probably the species was present in Lake Ontario as a part of its original range. Its access to the upper four lakes was blocked until the Welland Canal and the Erie Canal provided routes for it. Recall that the construction of these canals occurred about 1825 to 1835.

The alewife was identified as present in some of the upper Great Lakes by the 1930s. Their period of superabundance and their pronounced nuisance began in the mid-1950s. When we began to announce our intent to introduce salmon, the alewife was present in enormous numbers in Lakes Michigan and Huron, and present, though not as abundant, in Lake Superior.[5]

Biologists usually explain that alewife reached these levels of superabundance in Lakes Michigan and Huron because the lake trout had been reduced to near extinction by overfishing and by sea lamprey predation. The absence of an efficient predator is certainly part of the explanation, but I've always doubted that even if the lake trout had been present at pre–World War II population levels, they would have been capable of limiting alewife abundance.

Out of Balance

In addition to being a severe nuisance on the beach, other incidents drove this fact home to me. After I arrived in Michigan, I flew with the director to meet employees in the Upper Peninsula. The pilot mentioned that, since it was a very calm day, we might be able to observe schools of alewife in Lake Michigan. And we did! The size of those schools was absolutely unbelievable. We would estimate that we could see schools of alewife five to seven miles long, huge masses of fish food, definitely a **critical factor** for us.

At another time, when flying over Lake Michigan, I saw large, floating masses of dead alewives. The presence of dead alewives on the beaches and in the water of our swimming areas focused the concern on the problem that the species' abundance presented. Being a short-lived fish and with little or no pressure of a predatory species to limit its numbers, it would typically live its life span and then die. The most stressful time in its annual life cycle would be in late spring and midsummer. The alewife would be stressed by the long, cold winters and, following that period, would encounter the demands for energy in the migration to spawning areas and in reproduction.

So it was that the principal period of alewife die-off occurred in the middle of our beach-oriented tourist season. The complaints were emphatic. Resort communities lost their business. The beaches were described as a pile of dead fish a foot high and three hundred miles long, extending from Chicago to Mackinaw City. Another problem was that dead alewives repeatedly blocked Chicago's drinking-water system intakes.

The public outcry to do something was loud and clear. The Great Lakes Fishery Commission, through its action arm—the U.S. Bureau of Commercial Fisheries— was the management agency in control up until we changed that in Michigan waters. Their approach to solving the alewife problem clearly represented the thinking of those oriented towards commercial fishing. Their first efforts were to equip some of the Lake Michigan fishing boats with a trawl. These boats really were never designed for trawling, but they could pull a small one.

There was no problem in catching alewives.[6] The fishermen could catch alewives to a point where their boats almost sank. The problem was in finding a market. They attempted to market them as they were on the East Coast, and this could be done. However, it cost more to catch the fish in the Great Lakes than could be realized from selling them as fish meal and fish oil.

The U.S. Bureau of Commercial Fisheries had one other solution: to purchase six or seven of the largest and newest beach sweepers. These choices of response to the huge abundance of alewife clearly showed a lack of imagination, and I scoffed at them. We—the biologists of the Michigan Fish Division—saw clearly that this was not a problem, but an opportunity in disguise—an enormous food supply upon which to base an attractive and even world-class salmon fishery.

Prey

Some newspaper writers then and now assert that our principal intent in introducing salmon was as a means of controlling alewives. That was not the case. We announced early and often that we were going to build a world-class fishery for sport anglers. Yes, we were going to utilize alewives and, yes, eventually we might achieve a level of control. However, remember that we were fisheries biologists functioning as managers. We were interested in clean beaches, but that was neither our assignment nor our goal.

In dwelling on this point of argument, I've often searched for a comparison that would bring understanding. It was a little bit like saying to people who were basically in the business of raising cattle that they had been presented with an outstanding range of lush grass extending three hundred miles long and seventy miles wide. As they looked at it, do you think they would've said we need to put cows out there to keep the grass short?

Keep these ideas in mind as the salmon story unfolds.

Salmon

The Great Lakes salmon story didn't begin with our introduction of coho or chinook. The introduction of salmon into nonindigenous waters had a rather poor history behind it. Pacific salmon introductions had occurred in the Great Lakes at least thirty-five times before, and almost all were total failures.[1] Most of the failures could be attributed to the size (life stage) of the fish being planted, or the locations (habitat) where they were introduced, or the timing of the stocking.

Six separate species generally appear in the group under the heading of Pacific salmon. Of these, five—coho, chinook, kokanee, pink, and chum—appear on the eastern or North American shores and streams of the Pacific Ocean.

We had two principal areas of concern about introducing anadromous salmon into fresh water. First, would they likely be successful?

From the first days following the announcement of our intent to introduce coho salmon into the Great Lakes, we had supporters, opponents, and a third group that I would label the doubters. Our planned introduction simply stated that we expected coho to complete a life cycle without ever entering salt water, and to provide an exciting fishery in the process. Some people said it had never been done. We needed to respond, and we had a few seed examples to describe.

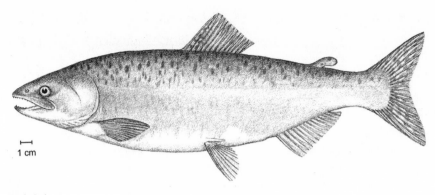

Pink Salmon

I have mentioned earlier that I knew of two freshwater instances—one in California and one in Montana—where coho had been introduced and survived to grow to maturity. There were still doubters and nonsupporters, and it was fortunate that we could offer yet another example of successful salmon introduction right here in the Great Lakes.

Pink Salmon

The pink salmon is one of the smaller Pacific salmon, reaching maturity in two years and spending a brief period in fresh water.[2] They appeared less attractive and, in the early days, had little value as sport species, but they proved to be a **critical factor** in our process.

Sometime in the early 1950s, the government of Canada began rearing pink salmon in a fish hatchery located on a tributary stream to Lake Nipigon, which has an outlet stream that flows into Lake Superior. The reported purpose of this program was to establish populations of pink salmon in a stream tributary to Hudson Bay to augment the food supplies of the native peoples in that area.

Salmon eggs were collected in the West and transported to the northern Ontario hatchery. By all accounts, the plan was successful up to a point. The egg-hatched fish were reared to the desired size and, from that point, were airlifted to the release sites in the target streams. To the best of my knowledge, those fish released in Hudson Bay streams were never seen or heard of again.[3]

Now you will get the true story of how pink salmon arrived in the Great Lakes in about 1955. Possibly as late as 1959, one of them turned up in a fisherman's catch from a Minnesota stream that flows into Lake Superior. It was recognized as a different species and, in due time, was sent to biologists at the University of Minnesota, where it was promptly identified as a pink salmon. How did this come to pass?

A fictitious explanation emerged. It seems that as the fish were being loaded into the float planes headed for stocking in Hudson Bay tributaries, one or more of the twenty-gallon milk cans being used as containers was accidentally knocked off the dock, and perhaps a few hundred or even a couple of thousand pink salmon escaped. That was the generally accepted explanation, although it seemed quite incredible to most of us that it could have happened that way.

In any event, pink salmon gradually spread, first through Lake Superior, spawning in many different tributary streams on its north and south shores. Eventually, pink salmon also became quite common in the northern portions of Lakes Michigan and Huron. Scattered examples of pink salmon have since been reported elsewhere throughout the Great Lakes.

Years later, a more credible explanation developed. In what has been described as a deathbed confession, the former supervisor of that initial hatchery admitted that he had released as many as twenty-five thousand young pinks simply out of curiosity to see what would happen.

The abundance of pink salmon has subsided in recent years, but they continue to exist as a self-sustaining species, at least through Lake Superior and its tributaries, as well as in Lakes Michigan and Huron. The significance to my salmon story, however, is that it was supporting evidence that we were justified in being optimistic in planning to introduce coho salmon. This example greatly aided our ability to convince decision-makers of the viability of our attempt to introduce coho salmon beginning in 1964.

Kokanee Salmon

When interviewed for the job as chief of fisheries for the Michigan Department of Conservation, I was asked if I might do anything special if I were selected. At the time, I said something about introducing kokanee salmon or at least considering it. It was very logical because there was extensive experience with this species in

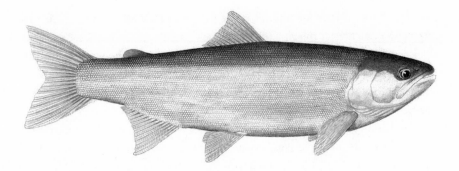

Kokanee Salmon

PETER THOMPSON, *FRESHWATER GAME FISH OF NORTH AMERICA*—USED WITH PERMISSION

fresh water. To begin with, kokanee salmon is the name given to the landlocked variety of the sockeye salmon.[4] Before 1900, populations of naturally occurring landlocked kokanee were discovered in British Columbia and Alaska. Early fish culturists stocked kokanee salmon in several locations, principally in the western states but in a few places in the East as well. I happened onto some records of kokanee present in Connecticut.

Second, the kokanee has several other attributes that make it desirable. It is an excellent sport fish; it is delicious eating. It also principally eats very small food items throughout the water column. Since it existed naturally in freshwater environments, it attracted early and frequent attention.

It was in late September or early October of 1964, and I had been in my new position as chief of the Michigan Fish Division for about a month. Attempting to do something "spectacular," I began to explore the possibility. I made several calls to friends back in the Colorado department. Most of these went to Wayne Seaman, chief of fisheries management. He said that we could get one million kokanee salmon eggs from Colorado if we sent a crew to work with their personnel to take the eggs from the spawning fish. If we were to do this, it probably would be in the latter half of October.

After some discussion with my staff and with fisheries professors at Michigan State, we decided to go ahead. We perceived that we could introduce the kokanee into a certain type of inland lake. We would pick some of the large cool or cold lakes. Our concept was that the kokanee would be added as one more attractive dimension of a fishable population for Michigan anglers, but in no way did we consider stocking it in the Great Lakes. I went forward with the idea to deputy director Chuck Harris, who remembered my comments from the examining interview. I quickly got the go-ahead.[5]

When the time came in late October, I sent an experienced crew by air to Colorado. Bill Mullendore, in charge of our publicity, organized our news media outreach, and he planned to have pictures and text sent to the press when the eggs were received here in Michigan. All the newspapers covered it, complete with lots of pictures. The egg take went well, and very shortly we had a million kokanee eggs in our hatcheries. More media coverage occurred throughout the incubation period.

The eggs were cared for, and the resulting young salmon were to be released in tributaries of selected larger, colder inland lakes in the spring of 1965. The lakes that we considered were Torch and Elk Lakes in Antrim County, Crystal Lake in Benzie County, and Higgins Lake in Roscommon County. During the busy months of early 1965, the news media, guided by Bill Mullendore, gave a lot of attention to the upcoming planting of kokanee.

The first planting was quite memorable. I had chosen the site—Wilkerson Creek, a small tributary stream to Torch Lake. Director MacMullan and probably, again, Bill Mullendore had convinced Governor Romney to plant the first of these fish. I went up a day or two before to be present at this event and to say a few well-chosen words for the public gathering and the news media. The governor flew in by plane and then by helicopter. We had prepared some statements for him. I joined him in the helicopter, and he briefly reviewed the prepared material and was ready for his role in planting the salmon.

Perhaps one hundred people crowded the banks of the small stream. Governor Romney made his way down to the water, turned, and gave a short series of comments expressing his hopes for success. He was handed a bucket containing water and a few small salmon and, with a flourish, bent over and released the fish into the stream. All this was caught on many cameras, of course.

Now Torch Lake is only about four or five miles from my childhood home in Bellaire, Michigan, and my parents were still living there. Of course, it gave me great pleasure to have orchestrated this event and to have them there. My mother got to meet the governor and was photographed with him.

Beyond the initial planting, the plan was that we would be able to take eggs again from Colorado for year two and then be independent with an egg source from our own lakes. This move to introduce kokanee was quite safe. It was generally known that many states either had programs of introduced kokanee or had at least attempted them.

Michigan governor
George Romney
inspects kokanee
salmon smolts before
stocking them in Torch
Lake.
MICHIGAN DEPARTMENT OF CONSERVATION/
NATURAL RESOURCES

After all these years, a couple of thoughts come to mind. I suppose I had been grandstanding a bit, but the stocking of kokanee salmon in Torch Lake was a complete failure. To the best of my knowledge, none were ever seen again. I don't know why, probably an inadequate supply of plankton. Another thought is that, if they had succeeded, the very small tributary streams entering Torch Lake were inadequate.

The other half of the kokanee salmon plant was in Higgins Lake. There they grew to maturity, and for several years, our department had a modest program of collecting eggs and releasing them in several other promising lakes. Soon we did no further stocking of this species.

Although it was high profile and, from my perspective, approached "spectacular," the kokanee introduction per se was never that important. It wasn't planned for this reason, but looking back I'm sure that the kokanee salmon program was an icebreaker of sorts. It naturally took a remote back seat as other, truly spectacular developments occurred.

That left two salmon species that would seem most likely to adapt to Great Lakes environments—the chinook and the coho. At this point, I need to mention some divergence in the common names used to designate these two species. On the West Coast the coho is often called the silver salmon, and the chinook is referred to as the king salmon. The American Fisheries Society—the professional organization that sets standards for a variety of rules for terms used in scientific writing—has a set of preferred common names. Since we were introducing salmon to new audiences, we chose to follow the list of preferred common names as established by the American Fisheries Society.[6] These common names are the chinook and coho.

Chinook

Michigan's Department of Natural Resources provides the following description of the species:

The chinook salmon is also a member of the Salmonidae family, which includes the salmon, trout and whitefishes. All these are characterized by an adipose fin (the small dorsal skin flap just in front of the tail), and prefer cold water with high oxygen content, making the Great Lakes an ideal habitat.

Chinook salmon reproduce by spawning in Great Lakes tributary streams varying in size from small tributaries to large-connecting waters. During the fall months (mainly September and October), chinook spawn ascend the streams from the big lakes, and then spawn over beds of gravel, typically on riffles. Within a few weeks after spawning, adult chinook die.

The following spring, chinook salmon eggs hatch and the young usually remain in the river for a few months before migrating out to the lake. In some cases, chinook smolts have been reported to spend up to 1.5 years in the stream before they migrate. Young chinook smolts in rivers eat insects, insect larvae and crustaceans; whereas adults in the open lake eat fish almost exclusively. In the Great Lakes, alewives make up most of their diet.

Once in the lake, males tend to remain for 1 to 2 years and females for 2 to 4 years before returning to spawn. Adult chinook salmon average approximately 15 pounds, but can reach 20 to 25 or even upwards of 30 pounds. Chinook salmon live in the open waters, also termed the pelagic zone, during most of their life span.

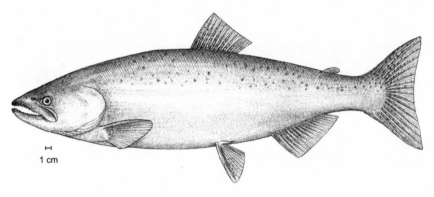

1 cm

Chinook Salmon

AN ATLAS OF MICHIGAN FISHES, USED WITH PERMISSION FROM THE UNIVERSITY OF MICHIGAN MUSEUM OF ZOOLOGY

In the late fall to early winter, they move into the southern reaches of each of the lakes, traveling many miles offshore as they go.

In the spring, they first move up the coast, then they move offshore depending on prey fish patterns and lake temperatures. By fall, mature chinook salmon will congregate at the mouth of their natal stream and begin their journey upstream to spawn.

Anglers prize chinook because of their large size and the challenge they present for fishing, and because they make a delicious meal.[7]

The chinook has an advantage over coho in that it is easier to rear and matures more quickly, putting less pressure on the hatchery system. However, we didn't have access to a supply of chinook eggs when coho became available.

Coho

In a December 1964 Department of Conservation *News Bulletin*, we provided the following description of the coho, with deliberate use of the director's choice word.[8]

It is a spectacular game fish, similar to the steelhead which now roams the Great Lakes and migrates up connecting streams to spawn. It puts up tremendous, leaping fights, particularly if hooked on light tackle, according to reports from western fishermen.

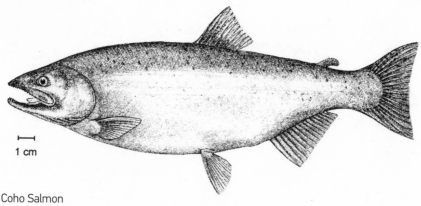

1 cm

Coho Salmon

In Michigan, this member of the salmon family is expected to provide best action right before and during its spawning runs.

In late summer, the fish will begin to congregate for these runs by coming into bays and other shore areas. During this time of the year, they will be taken by shallow trolling with flies, plugs, and other surface tackle.

Later in the season—probably sometime in October—Coho [*sic*] salmon will enter spawning streams and seek out gravel areas to give anglers a real go-around at sport fishing. These fish will continue to be good eating catches until they start to darken, a body change which takes place in preparation for spawning.

During other months of the year, troll fishermen will have to go after Coho salmon in deeper waters of the Great Lakes.

Timing of Coho spawning runs will supplement later migrations of steelheads to provide a longer fishing season, stretching from late summer almost continuously through the next spring in shallow, offshore areas and Great Lakes streams. . . .

The Coho is a short-lived fish. It dies after spawning once, usually in its third year. An exceptionally fast-grower during its last year or two, the ocean-going Coho has been known to gain as much as 10 pounds from May to late September during its third year.

In western streams, the average weight range for this fish is 6–12 pounds at maturity. Department fisheries men expect the Coho will run smaller in Michigan's fresh waters, probably between three and six pounds.

In appearance, as well as in its fighting manner, the Coho is quite similar to Michigan's lake-run steelhead (rainbow) trout. This silvery, fat-bodied fish has a somewhat blunted head like the steelhead. Its flesh is red and is said to be a real taste treat.

A plankton feeder at a young age, the Coho shifts its diet to fish during its second or third year. It is expected to feed chiefly on smelt and alewife in the Great Lakes. Populations of alewife, a low-value fish, have exploded in these waters during recent years to offer an abundant food supply for Coho salmon.

Introducing
Coho

The Call

had been in my new position as chief of the Fish Division for less than two months. I had been making the rounds and meeting with staff members and others interested in the division's work. I had done what I thought possible to meet Director MacMullan's "spectacular" challenge by introducing kokanee salmon.

Then, early one evening in late October of 1964, I received a phone call that would change my life. It was certainly the most important phone call that I have ever received. The caller was Dr. Robert Ball, my major professor and mentor throughout my graduate program at Michigan State College (now University). He was on the West Coast attending the annual meeting of the American Society of Limnology and Oceanography.

"Did you know that Oregon and Washington have a surplus of coho salmon eggs?" he asked.

Not only had I not heard that news, I couldn't believe it was true. It was a total reversal of anything I had ever heard. I was skeptical because I knew that the states that shared the Columbia River Basin had a pact that no salmon eggs would be sent anywhere outside of the basin. While I was still chief of fisheries research in Colorado, we had tried to get some coho eggs and had been turned down repeatedly, with this pact being cited.

Of course, I remembered working with Dick Klein in Colorado, who had succeeded in getting 150,000 eyed coho eggs from a private fish hatchery in 1962. They were stocked in Granby Reservoir, the largest body of water in Colorado (at that time). So, we had introduced a few in Colorado, and Montana and California had had successful examples of coho spending a full life cycle in fresh water.

I would like to dramatize that very special moment after the call. I would like to tell you that, as I considered the decisions that were ahead of me, I was standing on a bluff overlooking the vast waters of Lake Michigan. I was, but only in my mind. For most of the night, I sat alone at home in my big easy chair. There was no point in going to bed.

I began to consider the practicality of introducing salmon—coho salmon—into the Great Lakes. The fit was so perfect, the opportunities so huge, the consequences so monumental and predictably so controversial. I sat up all night, napping only for brief periods. I tried to tell myself, "Wait 'til morning to call the biologists in Oregon and Washington and confirm the news that they might have a surplus of coho salmon eggs." I remembered the attempts we had made while in Colorado, always to be told that no eggs were available.

How would these salmon fit biologically? I was sure the answer was "very well." The wisdom of John Hannah probably came to mind: "If you want to raise something, go where it will grow." Would the introduction of salmon, an exotic species, be good for the Great Lakes? My answer to myself was "yes," as the invasive alewife would be an enormous food supply. Could we expect the salmon to spawn successfully? Yes, Michigan had many streams that would probably be suitable. Yes, it would be a great sport fishery. I don't pretend that all these thoughts came in a logical sequence, but they came over and over. I cannot forget my excitement during those first few hours.

At the time, I thought only in terms of Lake Michigan. My own research background was focused on small inland lakes. This lake was huge—three hundred miles long, seventy miles across. Freshwater biologists had never dealt anywhere with lakes of this size, and it was only one of five Great Lakes. I thought mainly of biology, but what about spawning areas? Yes, we had spawning areas—most of the streams tributary to Lake Michigan, from the St. Joe to the Jordan at Charlevoix, including the Pere Marquette, of course. In the Upper Peninsula, the Menominee and some others would be good. They were suitable for trout and very likely suitable for salmon. Could we even get those salmon? What would be the impact? These were just shreds of thoughts that would be expanded and improved on, criticized, verified in the days and weeks ahead.

I was so excited! As I sit in my chair some fifty years later, I still feel an echo of excitement as those thoughts surface—a night view of the fishing scene, the thrill and some apprehension about what lay ahead is so difficult for me to describe adequately, not to embellish, but just to recall.

A New Day Dawns

had to wait, because of the time zone difference, until late morning to call the West Coast. Then I started contacting biologists whom I knew—Ernie Jefferies in Oregon and Cliff Millenbach in Washington. One **critical factor** that I had brought with me from Colorado was a working acquaintanceship with them. I had been a member of a professional organization known as the Pacific Fishery Biologists. I had attended several meetings and had met them there. I didn't go there to learn about salmon; I went to share experiences with the problems of water diversion's effects on fish populations. They were so many years ahead of those of us who were working with those problems in Colorado.

When I was ready to request eggs from Oregon and Washington, Ernie and Cliff confirmed that the report that I'd heard was true, and they helped open several doors—more pieces of the mosaic that made it possible for us to introduce Pacific salmon into Michigan waters. One of those pieces and another **critical factor** was the Oregon Moist Pellet.[1]

The saturation of the Columbia River system with dams and other elements of massive hydroelectricity generation nearly eliminated historic and marvelous runs of Pacific salmon species en route to their traditional spawning sites. In response to the demise of these wonderful and valuable fish species, the federal government, the states of Oregon and Washington, and the Bonneville Power Authority generated funds and efforts of many kinds to alleviate this destruction and to restore the salmon populations to health. Their efforts included constructing and operating numerous, very modern facilities for taking eggs, rearing them in hatcheries, and releasing them in large numbers. This effort and the research to support it began soon after World War II. The states developed a compact that they would not fill any requests for salmon or salmon eggs from outside of the Columbia River basin. They simply didn't have enough for their own needs.

For more than a decade no significant restoration occurred. "Smolt" is the name given to a young salmon as it begins the changes that prepare it for migration from fresh water to salt, and the smolts released from the hatcheries during that period simply lacked the stamina to survive the rigors of their environment and return to spawn. The states were chronically short of eggs. Hence, the basic refusal to ship any eggs elsewhere.

A Different Diet

In the late 1950s, however, much research and experimentation produced a new diet for feeding the salmon. It was called the Oregon Moist Pellet, and it produced stronger, more vigorous smolts. Their survival and growth was much better than fish raised on any previous diet. The first mature adult salmon reared on the Oregon Moist Pellet diet began to appear in the spawning runs of the early 1960s. For the first time, the agencies operating the hatcheries were getting sufficient numbers of eggs to sustain the maximum capacity of their salmon rebuilding programs.

It was the increased stamina of the salmon reared on the Oregon Moist Pellet diet that ultimately produced surpluses of returning adults at the very time—the fall of 1964—that they received our request for eggs. And because, and only because, they had surpluses did they approve our request, certainly a **critical factor** for us.

Gearing Up

I shared the news with my Michigan colleagues, and we began exciting conversations, as we would the next morning and the next and the next and the next. Of course, we had read the McFadden report, which recommended establishing additional Great Lakes areas, closed to commercial fishing, to enhance sport fishing, a strategy that the Conservation Commission had approved. But we had not taken any specific steps in that direction. The only thing we'd been able to do thus far was get some kokanee salmon from Colorado. That was easy. They were destined for inland lakes. But Pacific salmon were a whole different story.

We hadn't yet discussed what we would need in the way of fish species to create a sport fishery that would be so attractive that people would commit the time, the money, and the effort necessary to participate. We would need a large fish, most likely a member of the trout and salmon family, because that was the nature of the environment and it was also, from a social standpoint, near the top of most Michigan fishermen's list.

Otherwise, we were totally unprepared to deal with this outstanding opportunity that would involve taking over the total management of Michigan waters of the Great Lakes. We were not ready to do that. We had a lot of things to consider and to bring into that order of sequential development, but we kept talking biology. Almost assuredly, much of the coho diet involved anchovies and other similar-sized prey species in the Pacific. Would we have to keep stocking coho from hatcheries every year? Perhaps, eventually, some could come from spawning. Water temperature would be no problem. We had the exciting example of pink salmon in Lake Superior.

Was there truly an obligatory reason for salmon to spend some time in salt water? We began to develop answers once we reviewed every scrap of scientific literature concerning the salmon of the West Coast. Then we settled down a bit and began thinking, "What do we have to do first?" We had to make a convincing argument to get permission to request the salmon eggs.

We as a staff began to assemble all the information we could to help us understand the situation and how we would build a response to the opportunity before us. Wayne Tody searched the literature, and anyone with any experience chipped in. A review of the research conducted by this division over the years showed only one paper that was germane to our proposed new management program. I recited my brief experience with coho introduced into Colorado. I recalled that a full generation of this species had been experimentally raised in a freshwater hatchery in California. Another experience that I remembered was hearing from the Montana department that they had released coho salmon into Georgetown Reservoir and that mature fish had developed.

It was clear to me that what we were about to do—that is, request the eggs—would change the biology of the Great Lakes, create a very large and valuable sport fishery, and substantially alter the commercial fishery that had existed for 150 years. I was truly overwhelmed by how everything in my life seemed to coalesce around that one point of momentous decision.

Clearly, these huge lakes, chock full of a forage fish such as the overabundant alewife, were ready-made for the introduction of a large, predaceous species. In fact, at the time I described it as "a ripe plum on a low branch." But now, with the salmon potentially available to us, it was time to move. Indeed, everything I had ever been taught, or had ever experienced, told me that it was the right thing to do—despite conventional wisdom and the opposition that would inevitably form.

Seizing the Opportunity

There was a serious time factor, and we had a very short period if we were going to request salmon eggs. The coho of the Columbia River system usually spawn during the last half of September and first part of October. All trout and salmon have similar stages of egg development. After fertilization, and in the first stages of development, the eggs are very tender and will die if shaken or bumped. They are termed "green eggs." When development has progressed to the stage when the young salmon's eye is visible, they are called eyed eggs and will tolerate moderate handling and bumping. It is only in this stage that fertilized eggs can be shipped. Total time from egg fertilization to hatching depends on water temperature, but, in practice, the total period of development will be between thirty and fifty days. The eyed-egg stage—the period suitable for shipping—is roughly the last half of the egg development period. In general, the eyed-egg stage cannot be expected to last more than thirty days.

Oregon and Washington biologists had already taken the coho eggs for their hatchery program. For the first time, they had, at least in Oregon, retrieved a quantity of eggs that exceeded the number they needed to optimize their hatchery rearing potential. Of course, if they had known ahead of time that this would happen, they might have been better prepared to dispose of them.

In all likelihood, any eggs that would be available to us had already been taken from the fish, and those eggs would have been fertilized within minutes of collection. In other words, the clock was ticking! We had no idea how we could clear all the hurdles to gain approval at our end in such a short time, and the Oregon and Washington biologists had to clear similar hurdles at their end.

It would all have to occur somehow by the middle of December! Those who have worked in a state bureaucracy know that accomplishing things takes time. There was so much to be done, so many questions to be answered, so many possible roadblocks to overcome. There was much to be done and little time to do it. Sometimes it seemed impossible. We made many phone calls to our counterparts in Oregon and Washington. Yes, the eggs were available. But in those fish divisions, they needed the approval of their superiors. They couldn't guarantee that, but they would try.

Supporting and Opposing Forces

We knew that we would have to have significant support to undertake this initiative. We also recognized that we were likely to encounter opposition. We committed ourselves to making the best of both.

Administrative Approval

Assembling all the information we could, Wayne and I went to Deputy Director Harris. He was interested. He sensed our excitement and, I think, grasped the immensity of the project that we proposed. His reaction was to schedule a second meeting, this one with Director MacMullan. The director's reaction was similar and practical. At the November meeting of the Conservation Commission—the policymaking board of our department—he included a legislative proposal to extend the commission's authority to regulate sport fishing in the Great Lakes. This would complement the department's existing authority to regulate commercial fishing.[1] It would build upon the recommendations of the McFadden report, which the commission had approved in the summer.

Over the following months, we developed and proposed various measures to limit or eliminate commercial fishing for key potential sport species, all while obtaining and rearing coho salmon for introduction into the Great Lakes. How very fortunate we were to have a positive response from the leadership of our department, to have a front office geared for action, and to have the circumstances needed to proceed with our introduction of salmon. This was definitely a **critical factor.**

I am reminded again of Tom Brokaw and his descriptions of the "greatest generation." Director MacMullan had served as a bomber pilot over Europe. Deputy director Warren Shapton had served as a naval officer (in the North Atlantic, I think), and deputy director Chuck Harris—my immediate supervisor—had served as a Marine major and had led Marines ashore on the bloody island of Tarawa. The director and the deputy director had been in those positions less than a year, and it was so extremely important that they were men of vision and decisiveness. The majority of the leaders in the sweeping changes that were occurring across the State of Michigan were members of the "greatest generation." At the time, I didn't think of it, but with hindsight I firmly believe that was an important factor. Their leadership was critical when we had to move quickly through the established administrative process to secure, in the window of time available, approval for the introduction of Pacific salmon. They had the talent, the training to understand the components of our plan and to judge them as well worth a try. Their approval was quick, firm, and sustained.

Making It Official

I was authorized to make an official request to both Oregon and Washington for one million coho salmon eggs from each state, and I did that as soon as possible. After considerable discussion, their answers were tentative yeses. They told me that they would get back to me as soon as they could. Within a day or two, both departments had a more definitive answer. They said that they would send us one million eggs apiece if we made a formal request. This turn of events hit the division like a bombshell!

We sent our requests by letter, and unfortunately, the State of Washington formally declined their assistant director saying that they did not, after all, have surplus eggs, even with a good run.[2] They also expressed doubts that developing landlocked reproducing silver (coho) salmon would be possible. However, they would reconsider in the fall of 1965, and they did contribute eggs later that year.

Oregon replied more affirmatively, saying that they were agreeing to send us five hundred thousand eggs.[3] Their letter went on to stipulate certain conditions. These conditions were totally acceptable to us and really constituted the only advice we got. One of these was that we must agree to rear the fish on the Oregon Moist Pellet diet. The second was that, when released, the salmon would be placed in large numbers at a few locations where we might expect the fish to return to a stream suitable for spawning, rather than scattered at many locations.

We were poised to make the great leap.[4] We had the "green light."

Beware of the "Exotic"

However, others were not so quickly persuaded. The most compelling argument that our initial critics made against us was that the coho was a nonindigenous fish and that such a species should never be introduced into the Great Lakes. If it were, it should be done only after very careful (and time-consuming) study of the possible consequences. This was the standard view among many fishery experts, and it was quite compelling, mainly because it was based on hard lessons of past experiences.

We had reckoned with the potential of public groups coming forward with the argument that we should not introduce an exotic species. We had plenty of examples before us of the troubles that had occurred when such species were introduced, and we knew that those would be thrown at us.

Between 1870 and 1910, many nonindigenous species were moved beyond their original home areas, with sometimes tragic results. The worst and most prominent disaster was the distribution of carp from Europe throughout most of the lower forty-eight states as part of a massive federal program. The result was a plethora of unwanted carp populations throughout a broad range of North America from northern Mexico to southern Canada. It was a huge mistake because the carp altered the ecosystems where it lived and either completely displaced or came to dominate truly valuable native species.[5]

Furthermore, our opponents could point to the entry of the sea lamprey into the upper Great Lakes to argue that only bad things happened with such introductions.[6] The sea lamprey was an obvious detriment to the fishery biota of the upper Great

Lakes, and we would quickly point out that no one had, or ever would, deliberately introduce sea lamprey.

Then they would charge, "What about the smelt?" which by then was recognized as deleterious to sustaining a lake trout population that spawned in the open waters of the Great Lakes. The American smelt had been introduced in about 1913 by a sportsmen's club with no scientific background.[7] That wasn't a fair comparison.

Well then, what about the alewife?[8] Again we pointed out that no one had deliberately allowed the entrance of alewife into the upper Great Lakes. That species had simply utilized the constructed canals as shallow entry channels.

Then we would point out: look worldwide where trout and salmon have been introduced from their original home in Europe and North America to every continent except Antarctica. Many of the early introductions had been from England, with its widespread colonies. English leaders, many of them avid fishermen for Atlantic salmon and European brown trout, went far and wide. Following the attempts to introduce one species or another across the globe are remnants of their successors: brown trout in the streams of the Himalayas, in Kenya and elsewhere in Africa, in Australia and New Zealand and South America, and, of course, in our country and Canada.

Less successful had been the introduction of Atlantic salmon, which had been tried in many places, and with only some very limited success in New Zealand. When we came to salmon from the northern Pacific, we learned that they had been introduced in many places, nearly always without success. On the South Island of New Zealand (incidentally, I fished this population after I retired) were some successes of stocking the various species of trout and salmon.

Our detractors were quick to point out documentation that the introduction of Pacific salmon into the Great Lakes had occurred at least thirty-five times before and that all were total failures.[9] According to John Parsons's report, *History of Salmon in the Great Lakes*, "Somehow, up to 1950, apparently the right species of the right size or age was never planted at the right place, at the right time, and in adequate numbers to establish either a temporary fishery or a permanent population." With the exception of pinks, only a few were ever seen as adults. Because the field of fisheries science had matured with a growing understanding of how to match fish species with their environment, we could point out that those introductions had occurred some fifty or sixty years earlier.

West Coast Assistance

Of course, much of the science related to salmon populations in general had been developed on the West Coast in response to major declines in their native environment. The **critical factor** in our favor was that the Oregon and Washington fisheries biologists had generously shared their knowledge and understanding and had provided important guidance. They recommended that we proceed with early steps of care of the eggs and the rearing of the salmon to a size of six or seven inches. We had direction as to where and when the salmon should be planted. It was with this background knowledge that we could proceed with optimism despite knowing that all previous attempts had failed.

A sizable number of biologists said that salmon was absolutely obligated to spend part of its life cycle in salt water. They would argue that once a salmon had smolted, it could breathe only in salt water. However, some of us knew that this was not so, and we regretted the fact that our critics were not better informed of the latest scientific understandings.

Understandable Caution

There were more failures, but more importantly and of concern to us was whether, if ever, the introduction of any salmon or trout had proved to be detrimental and produced a measurably undesirable result. We found two—only two—examples. One that I was particularly acquainted with was the introduction of the brook trout throughout the waters of the West, where only native cutthroat trout had been present. Yes, truly the net result of the brook trout introduction had to be judged unfavorable.[10]

The second was in Lake Titicaca in Bolivia and Peru of South America, where the rainbow trout (steelhead) had been introduced.[11] One or more small species of fish native to Lake Titicaca were the main catch of local fishermen and sustained many people as a food supply. The introduced steelhead nearly wiped out this indigenous species and proved undesirable or unwanted by the local people.

However, we were still faced with some very real questions, such as: How would coho salmon from the waters of the Pacific and the Columbia River basin do in the Great Lakes? Where would they go to spawn? Fortunately, fish managers from Washington and Oregon were very helpful in giving us a sense of the possible answers to those questions.

Good Neighbors

Besides the dismal history of fish introductions into nonindigenous waters, political factors were also going against us. For at least a couple of decades, the states of Minnesota, Wisconsin, and Michigan had had a tri-state cooperative program. We shared fisheries and attempted to work effectively together. In my personal history, the first report that I gave before a group of fishery professionals was presenting my master's fishery experiment results to these colleagues at the Higgins Lake training center, probably in the earliest months of 1948.

Among our long-standing, though not written, agreements with these states regarding fish management policies in waters that we shared was that none of the three would introduce a new species of fish without first obtaining the agreement of the other two. Thus, if we proceeded, we would be abrogating our agreement.

In the case of the coho, though, we couldn't possibly obtain the approval of the other two states in time. As we assessed our situation, we recognized that commercial fishing interests were still very strong in Wisconsin and Minnesota. We felt that it was quite likely that one or both of these states would object to the salmon introduction if we were to ask them. Since our decision to try to obtain coho eggs had to happen quickly, my decision was not to ask them. We decided that we would just have to deal with any objections later. We were going full speed ahead. Yes, I had chosen to violate that agreement and, in this case, I would plead that the end justified the means. We never did ask, and fortunately the issue was never pressed forward by the other two states, though their officials occasionally reminded us that we had broken our "gentlemen's agreement."

Great Lakes Fishery Commission (GLFC)

It was a plus that the GLFC had been conceived, funded, and established, complete with offices and staff. Their assigned task was to develop effective control of the sea lamprey. GLFC staff were conducting very effective research on how to control that parasitic fish species.

Population levels of many commercially important fish species were at an all-time low. The prevailing opinion then was that parasitic predation by sea lamprey was the cause. Another important effort was to restore lake trout populations by constructing some fish hatcheries to rear the species.[12] A new trout hatchery on the Jordan River was to be dedicated the following spring, and I was to be the department's representative on the speaker's platform. As our knowledge of the lakes and of the events capable of modifying fish populations increased, we realized that other factors were probably at work as well.

For example, during and following World War II, commercial fishing efforts increased dramatically.[13] New nylon nets were stronger, did not require drying (as had always been necessary with cotton and mercerized cotton nets), and, being made of much finer threads, were less visible to fish. Therefore, catch rates were much higher, which led to overharvesting. From a combination of causes, then, the Great Lakes commercial fisheries of the United States and Canada were in distress. The GLFC had been a response to this distress, and it had subsequently become the main policymaking group on issues related to Great Lakes fisheries.

Finally, the GLFC was charged with supporting the redevelopment of the commercial fishery. That explains why programs were devised to commercially harvest alewife and to develop more viable marketing strategies to improve the profitability of a variety of fish products.

When we began to actively pursue our management goal of establishing a sport fishery in Michigan's Great Lakes waters as the key value, we definitely faced opposition from the GLFC. The position of the commission in regard to our program of introducing salmon was generally negative, although seldom openly so.

A third-generation commercial fisherman named Claude VerDuin was one of the eight men on the commission, and his leadership skills quickly made him the main person to deal with in all matters coming before that body. VerDuin was no longer actively fishing by the mid-1960s,but was publisher of *The Fisherman*—a trade journal for Michigan's commercial fishing industry. Intelligent, polished, and politically savvy (one-time mayor of Grand Haven, Michigan), VerDuin was a very effective leader of commercial fishing in Michigan. From the first, he clearly understood that our plan would prohibit commercial fishermen from taking the species that would now be allocated to recreational fishing. Claude was always polite, even soft-spoken, but he was totally opposed to the planned introduction because he thought it would devastate commercial fishing on the Great Lakes.

The GLFC then included several other commissioners from both countries who were strongly associated with commercial fishing interests. In Canada, the staff assigned to the commission was from the Ministry of Fisheries and Oceans. Programs of this ministry elsewhere were predominantly concerned with commercial fishing. In other words, the deck was stacked against us.

The opposition of the GLFC to our salmon program found voice in various writers and in critical comments and questions from various sources, including the Michigan State Legislature. Its opposition was to be expected. The GLFC was now cast by events to be "defenders of the status quo," while we in the Fish Division were in the role of the "change seekers."

Interestingly, I was frequently required to attend meetings of the GLFC, as were the fish chiefs of other states. I made at least one direct attempt to shift the position of the commission towards the changes we were seeking. In fact, in its own report—*A Prospectus for Investigations of the Great Lakes Fishery*—the commission seemed to acknowledge the growing importance of the sport fishery.[14] "Knowledge of the amount of the sport fishery production is becoming increasingly important for the sound management of the fishery. Recent estimates indicate that the annual catch of some species by anglers may exceed several million pounds and that the sport fishery is a major user of the resource in certain areas."

My speech to the commission clearly stated our intent—to replace the commercial fishing allocation with one of recreational angling based on the introduction of salmon. I asked them to make a similar shift and to align with our goals. The GLFC's response to this speech was, to say the least, perfunctory. After listening politely, the commission thanked me and moved on with its agenda. Not one comment; not one question. Stony silence. With the benefit of many years of hindsight, I think I can understand that for that commission, at that particular time, a change in direction was impossible.

U.S. Bureau of Commercial Fisheries

The U.S. Bureau of Commercial Fisheries employed the action personnel of the GLFC, and their opposition was much more visible and frequent. In public meetings and in the press, they made multiple criticisms. Some were subtle, while others were more direct. They basically said that our program would fail; that we were introducing an exotic species; that we should stick with the native lake trout; that we were putting commercial fishermen out of business; and that we had not studied the long-term results of our venture.

That last charge was true. We clearly were acting very quickly and "by the seat of our pants." Perhaps with some bias, though, I would say that, despite our haste, we always knew that we were right. We were using good judgment based on solid biological factors, and I felt that these factors were ones that all conservation professionals were capable of understanding. All, perhaps, except one outspoken and influential individual, who seriously underestimated us.

The opposition to our salmon introduction from the U.S. Bureau of Commercial Fisheries was usually expressed by the person in charge of the region—Dr. Fenton Carbine, a product of the University of Michigan fisheries program. I believe he worked for a time at the Institute of Fisheries Research (IFR). It was at the institute that I met him for the first time while I was a graduate student. That would have been in the late 1940s.

Our paths crossed again when I accepted my first full-time appointment as assistant to Dr. William Beckman, also from IFR. As leader of the Colorado Cooperative Fisheries Unit, Beckman was an employee of the U.S. Fish and Wildlife Service. His supervisor was in Washington, DC, and was the same Dr. Fenton Carbine. It was Dr. Carbine's decision to withdraw federal support of the Fisheries Co-op Unit and to move Dr. Beckman to Washington (I believe this was desired by Dr. Beckman). Anyway, Carbine created the opportunity for me to become leader of the Colorado Cooperative Fisheries Unit, where I functioned for the next nine years and gained considerable experience, including participation in programs of planting kokanee salmon and coho salmon.

One time, trying to find some common ground, I asked to meet with him. He agreed, but did not want me to come to his office. At his request, I met with him at his home in Ann Arbor. The meeting was strained and short. He basically said there was no common ground available to us, and, at least in this, he was correct.

During the time that we were seeking a special legislative appropriation of $500,000 to buy Oregon Moist Pellet from out of state, the bill had passed the state house and senate and was on Governor Romney's desk for his signature. At the same time, the governor received an official letter from Dr. Carbine citing his arguments against the coho planting program and asking the governor to veto the appropriation that we were seeking.

The governor's office had an established process that it used in developing responses to correspondence. The Carbine letter was sent to the office of the director of Conservation Department, and the director forwarded the letter to the division most germane to the subject of the letter that needed a response. In due course, then, Dr. Carbine's letter to Governor Romney came to my desk for a response. Obviously, he was not aware of our bureaucratic processes.

I shared this bombshell of a letter with Wayne Tody, and we discussed our response with Deputy Director Harris. This was clearly an instance of a federal official attempting to interfere in the state government of Michigan, so we decided to "leak" the letter. Scores of copies "somehow" made their way to legislators, sportsmen's groups, and the news media.

Although Dr. Carbine was censured by his superiors over this letter, he continued his opposition. He continued to contend that the introduction of salmon would never control the vast overpopulation of alewife in the Great Lakes—one of the rationales we were using to push for coho introduction. Even after the first year of coho fishing was a great success, he termed it a "flash in the pan." Unfortunately, this did not end Dr. Carbine's attempts to interfere.

Trout Unlimited

In sharp contrast to the perspective of the Great Lakes Fishery Commission and practices of the U.S. Bureau of Commercial Fisheries, the concept of transforming the beleaguered Great Lakes fishery from commercially oriented to sport-focused also attracted supporters, and Trout Unlimited (TU) was a notable one.

TU is a nongovernmental, not-for-profit sportsmen's group, started in Michigan based on conservation principles. TU and its affiliated groups seek to maximize the recreational aspects of the fishing experience while attempting to protect its sustainability, along with the associated lakes, rivers, and streams, and surrounding land.[15]

Pacific Salmon

Trout Unlimited interacted in many ways with our Pacific salmon effort and the fisheries in general, some of them visible and some not so visible. There were financial implications and there was influence with the governor. For example, one early member—Alvin MacAuley, a banker from Detroit—was known to call Governor Romney at home or elsewhere at any time. He influenced the budget, and he and others influenced those who might become Conservation Commission members. Trout Unlimited was important to the budget process throughout my years in government, so I will tell some of their early stories that relate to me and to the Great Lakes salmon story.

Catch and Release

One of my projects in Colorado measured the difference between the survival of trout caught on live bait and those caught on artificial flies. The brief article that I published with one of my students presented data that showed a much heavier mortality by trout caught on live bait. The significance of this was the growing movement among some anglers to encourage a program that came to be called "Catch and Release." The logic was simple; after all, the primary elements of a sport fishing trip had been completed when you successfully caught a fish. That fish had, by then, provided most of the recreation that it could. There wasn't any logic that said you should kill that fish unless, perhaps, you were hungry. This small article caught the attention of Trout Unlimited president George Griffith and other early members.[16] It provided documentation of the logic in urging regulations that call for stretches of streams to be designated "flies only."

The information was timely. Trout Unlimited was going to have its first national meeting in Butte, Montana, and I was asked to speak there.[17] The ultimate significance of this was that I would travel from Michigan to Montana with George Griffith and other prominent members of Michigan Trout Unlimited and establish friendships and useful working relationships.

Sometime in the spring of 1965, George came to the Conservation Commission meeting. He often attended these meetings, having been a commissioner himself and a member of the previously mentioned committee that investigated the conservation department and urged modifications. In this instance, George came to the commission saying that he was concerned about northern pike coming out of the Mio Reservoir on the main stem of the Au Sable River, running up the South Branch, and being a serious predator on the brown trout, principally in the area of the Mason Tract. I then organized a trip to look for northern pike in that area. We saw everything from bank to bank using an underwater light, but never saw a northern pike.

The object of my story? George was very appreciative of my effort and an enduring friendship was established. Occasionally, George and I would spend some time together. My wife and I visited George and his first wife, staying at their guest cottage—the well-known "Barbless Hook." He loved that stream and fought hard and successfully to promote the goals of Trout Unlimited.

Mason Tract

In the early days of Trout Unlimited, a fairly large group of anglers began to enjoy fly fishing on the Au Sable. Many, including George Mason, desired to set aside the South Branch for fly fishing only. Several of them had purchased quarter sections of land to consolidate the stretch into combined ownership.

As the story goes, they met and played poker to see who might acquire or collect all the diversified ownership into one. At the end of the game, George Mason was the winner and acquired most of the scattered pieces in this manner. Prior to Mason's death, he and George Griffith worked out the details of transferring the ownership to the Department of Conservation with stipulations that it never be sold, and so it was when I came back to the state as chief of fisheries.[18]

However, a subsequent and more accurate survey and mapping revealed that a couple of parcels of land not owned by Mason and not owned by the conservation department had been missed in the early arrangements. This would have made it possible for other owners to build streamside cottages on these locations and violate the integrity of the tract. The Fish Division had a long-established program of acquiring key high-priority streamside and lakeside properties. In due course, these parcels were acquired and the Mason Tract was made whole.

Flies Only

A significant percentage of the early influential members of Michigan Trout Unlimited were from Grand Rapids. Many, but not all, were members of the Indian Club on the Little Manistee River, or the Watershed Club on the Upper Manistee River. The Indian Club owned several miles on both banks of the Little Manistee River, so the property was private. Under Michigan law, anglers are permitted to fish navigable rivers wherever they can achieve legal access. Translated, this meant that the public could fish those several miles of stream within the property of the Indian Club only by entering the stream at the upper or lower end. This effectively limited most fishing efforts to members of the club.[19]

Prominent in the membership of the Indian Club were three men and their families. They were Cornelius Shrems, Blake Forslund, and Robert Evenson. The club, led by these three men, came to me in the Fish Division saying words to the effect that if we would establish a "flies only" regulation on that portion of the stream, they would make their property open to anglers who fished with flies only. Their offer included the erection of stairs over their fences. We began to establish these regulations. A part of this process would be public hearings, which leads me to another story.

The concept of establishing sections of public streams limited to the use of artificial flies only was completely new. Many local people would be opposed. A public hearing was arranged in the town hall at Wellston, Michigan. I was to be the principal speaker to explain the new regulations, their purpose, and the cooperation that the Indian Club would provide.

It was an evening in late May. The weather was warm and very humid. I arrived a bit late, to be met in the parking lot with this scene. The lot was full of cars. A small, old-fashioned town hall had all the windows open, and it was packed with people. The sound of a lot of talk, some of it loud and a little angry, was pouring out of the windows. O. J. Bennett, the regional chief of our Law Enforcement Division, dressed in plain clothes, greeted me as I emerged from my vehicle. With real concern, he said, "Howard, that's an angry crowd in there. If this meeting gets out of hand and the crowd gets dangerous, I want you to know that I have seven plainclothes officers in the front two rows, and their task is to get you out the back door and into a vehicle that we have waiting there." Well, thanks a lot; that really made me feel good.

This was the most memorable test of my eloquence and any ability that I had to be persuasive. Yes, the officers were there, and I recognized some of them. On the other side of the audience a row or two back, I recognized the friendly faces of most members of the Indian Club, including the three men mentioned earlier.

It was touch and go for a while, and there was a mixture of approval and disapproval. When questions and answers were called for, a drunk near the back of the room repeatedly rose to make the same statement, words to the effect that "the bridge should not be called the Indian Club Bridge, it should be called the Johnson Bridge, named after my ancestors who were here long before the Indian Club." I called on him each time he raised his hand, and he would repeat the same statement. Gradually, a sense of humor developed, the crowd quieted, and the officers had no need to help me escape through the back door.

I cherish this memory and the strong friendships that developed among us, partly as a result of that episode.

The Jordan River Valley

The Jordan River Valley was nearly all in public ownership. Since then, public ownership has been expanded and more holdings have been acquired. One day in the Fish Division office, word came to us of the pending sale of two or three quarter sections of land very near the headwaters of the Jordan River. These parcels were about to be sold to a person seeking to build and operate a private trout hatchery. Private hatcheries have often proved to be the source of spreading fish diseases in the waters to which they are connected. A private fish hatchery in that location would be a serious threat to the Jordan River, which was a very special place for me.

As it was explained to me, we had money available to acquire these parcels, but this kind of expenditure came under the heading of capital outlay. For the department to spend money for this purpose required legislative approval, typically, through a budget process that we would have to have anticipated and secured the previous year. Translated, what that meant was that we didn't have time enough to get the necessary approval to purchase those parcels and prevent their acquisition by people intending to build a trout hatchery.

However, there was one obvious course to manage this situation. I called Bob Evenson. At that time, Bob was the president (and probably the principal owner) of a company known as Michigan Wheel. When I explained the situation, he promised to get together with Blake and Corny and get back to me. It was perhaps twenty-four or thirty-six hours later when Bob called me. He said, "We own the property, and we'll sell it to the state and get our cost as soon as it is convenient." Problem solved, thanks again to members of Trout Unlimited. Today the whole upper river valley is in public ownership and is managed as a state forest.

Catchable Trout

As Wayne Tody and I planned for the necessary hatchery system in which to rear the salmon, we needed to close the long-established, expensive, wasteful, and questionable policy of planting of "catchable-sized" trout. Trout Unlimited had consistently pointed to the research evidence that this was a wasteful process and had very little merit or stature as a conservation measure. So, we closed down this program to accommodate our salmon plans.

Not only did we acquire the space needed for the salmon through this action, we also earned a tremendous amount of goodwill from the very influential Trout

Unlimited. The solid support gained by this and other actions would reward us in many ways. TU members would be effective supporters of the Fish Division budget requests and of the changes in trout stream management that I've already described.

Salmon Support

I want to tell one more story that illustrates the support from the early membership of Trout Unlimited. Casey Westell, a member of TU's founding group, was an executive in a company known then as American Boxboard, with offices and a manufacturing plant in Manistee. When the salmon fishery in Lake Michigan literally exploded in 1967, tens of thousands of anxious anglers were trying to get a boat in the water, and tens of thousands of salmon were being caught. There was one small launch facility on the northwestern shore of Manistee Lake at the mouth of the Manistee River near the center of the action. Fishermen with boats and trailers lined up and waited for hours for an opportunity to launch through this very limited facility. Casey organized the local Lions Club membership to handle the waiting, impatient anglers and to expedite, to the best of their ability, the launching and recovery of those small boats headed for the big waters of Lake Michigan. He and his group provided yeoman service to help the new salmon fishery succeed.

Today Trout Unlimited chapters exist across this nation and beyond. It is an extremely effective force for the protection and enhancement of trout populations everywhere. It is thriving in Michigan, with an office in Lansing and a competent professional fishery biologist—Dr. Bryan Burrows—as its executive director. It continues its sustained support and achievement in the protection of Michigan's cold-water resources.

Three Possible Outcomes

All of that said, as we faced the introduction of coho salmon into the Great Lakes, we considered essentially three possible results: First, it would fail and, while that would have meant that we had wasted a certain amount of money, that would not be a tragedy. Second, it might prove detrimental to native species, which probably would include principally the lake trout and perhaps the whitefish. If the net result were more negative than positive in that direction, we would be severely criticized. Third, the result that we wanted and predicted and aimed for was to create a sport fishery—a sport fishery of magnitude and value that would exceed any other freshwater sport fishery in the world. We could find no substantial reason to believe that a successful introduction would have a negative effect, and it seemed to be a near perfect fit biologically to produce a large, healthy, perhaps even self-sustaining population of salmon. There was simply no reason not to proceed.

Eggs from Oregon

O f course, we were disappointed when the chief of the Washington Fisheries Division notified us that they would not be able to send any eggs after all. Cliff Millenbach told us that a member of their legislature had heard about the proposed transfer of eggs and had proceeded to block it.

After receiving approval for five hundred thousand coho eggs from Oregon, I arranged to send two of our Fish Division staff members—Dr. Leonard Allison, our fishery pathologist, and Emerson Krieg—to meet with staff members there. Both were very experienced fish culturists, and they went west with instructions to learn all they could about what lay ahead of us in rearing the young fish and making decisions about release locations. While they were there, they learned that Oregon would, after all, provide five hundred thousand additional eggs to help make up for the lack of those from Washington. In due course, the eggs arrived from Oregon in great shape, and we were committed.[1] Bill Mullendore orchestrated lots of publicity about it, and reporters and outdoor writers from all major Michigan newspapers and sporting publications picked up the story.[2] The eggs were safely tucked away in Michigan trout hatcheries. We were poised on the threshold of a great adventure, and here's what we could see ahead of us. The period of waiting until the first eggs were reared, until they had reached the size of six or seven inches, would last all of

1965 and a portion of 1966 when the smolts would be stocked, timed to take place in early spring.

We would have to sustain the enthusiasm and expectations until that date, approximately sixteen months away, and then it would not be until late summer of 1967 that our fishermen could experience the presence of most of the fully grown adult coho salmon. We would have problems, obstacles, and deficiencies to overcome.

Planning in Progress

We sat down together to plan and to assess our situation. In numerous meetings with staff members and individual meetings between Wayne Tody and me, we first began to grasp the full magnitude of what we were undertaking and how little we had in the way of physical and human resources, particularly experience. Most of those early conversations were between Wayne Tody (soon to become Dr. Tody) and me. I had assigned Wayne the task of leading our planning efforts, and we would exchange statements such as the following:

- Do you realize what you've got us into?
- We don't know enough about these Great Lakes of ours. We haven't conducted a single research project!
- How many times have others tried to introduce coho salmon into fresh water and so few of them have succeeded?
- We have no idea how much survival we will have from smolt stage to adult maturity and spawning.
- We don't have a fraction of the hatchery space needed if we are dependent upon maintaining populations with little or no survival or natural reproduction.
- We don't have the money in the budget.
- No fishery biologist or fish division in the world has ever attempted to construct fish populations in waters of the size of our upper three Great Lakes at any time, anywhere.
- We don't have a single biologist on our staff that has any hands-on experience with salmon. No one on our staff has any experience on the Great Lakes either!

Despite these issues and unknowns, everyone within the Fish Division remained optimistic. The fit between the enormous food supply of alewife and the reputation that coho had on the West Coast as a voracious eater on similar small pelagic fishes was perfect. The water temperatures in our Great Lakes were comparable. We had streams that were probably suitable where we might eventually achieve some level of natural reproduction.

Natural Reproduction

As we began to assess our capability of adequately stocking the Great Lakes with salmon, the other aspect to be explored was how much natural reproduction we might expect from the stream areas available to anadromous fish—in this case, coho salmon. Again, we relied on established research information from Oregon and Washington. The question was phrased like this: "How many migrating smolts can we expect from a given portion of spawning stream?"

We hardly knew how to ask the question, but a general answer emerged as a rule of thumb, based on the guidance of our West Coast colleagues. It is reasonable to expect one smolt for every square yard of spawning gravel. We constructed a survey form for field personnel, particularly fishery biologists, to begin to develop an estimate of how many square yards of spawning gravel would be available for each of the upper three lakes in the accessible areas of the tributary streams. Most of our streams flowing into the Great Lakes were at one point or another blocked by hydroelectric dams. That would limit anadromous fish such as salmon to those stream areas below the dams and to any tributaries entering the mainstream.

Another question was "How much additional gravel for spawning would be available if dams were removed, fish ladders built, or fish were manually transported to upstream areas for a part of the program?" It would be several months before this was complete, but when the results were tallied, we just looked at each other. Natural spawning was highly desirable, and we would promote it to any reasonable extent. However, it would clearly, on the information given us from West Coast experience, remain totally inadequate to fully stock the upper lakes. Conclusion: hatchery production would remain important as far as we could predict.

Scientific Review

It was also a logical, important step to conduct a thorough review of existing scientific literature. It was true and obvious to everyone that there had been no time for specific experiments and research on the Great Lakes and their connecting waters per se. However, that did not mean that we knew nothing of the existing literature on salmon populations and all elements relating to that. I brought my experiences in Colorado involving both kokanee salmon and coho salmon. I also contributed a small portion of existing knowledge gained from the successful introductions and/ or holding of coho salmon in completely freshwater environments in Montana and California.[3]

Another source of expertise was the writings of Dr. William Ricker, part of the faculty at the University of British Columbia. He was extremely well known in the fisheries profession and had published numerous articles relating to various species of salmon.[4] He had focused especially on coho that had completed life cycles in freshwater Cultus Lake.

The most important source of published information related to work on the Columbia River system. The very extensive development of dams on this system during and following World War II had done serious damage to the fabulous populations of several species of salmon that had existed there in earlier times. Billions of dollars and many years of research had been expended on how to augment and to manage migratory salmon populations.[5]

Most Columbia River salmon runs were badly depleted up to and including early 1964. The Oregon Moist Pellet development had produced stronger, more vigorous, better-surviving artificially reared coho smolts. That was why, for the first time, salmon were available to meet our timely request from Michigan. This research data was available to us, and much of it related to hatcheries and, therefore, was readily transferable for us to use in that portion of our program.

An oft-quoted statement from Dr. John Hannah comes to mind. In his speeches, when he laid out a series of problems, potential solutions, and opportunities, he would typically close his remarks with the following statement: "Yes, there is much to do, but there is also much to do it with, so let's get on with it."

The Hatchery Situation

For all that we understood about the disproportionate role that hatcheries had played in the Fish Division, we also realized that we needed a hatchery system capable of supporting our introduction of Pacific salmon into Michigan waters of the Great Lakes. It quickly became apparent that we had some problems to deal with in the existing state hatcheries.

Applying the Oregon Experience

What did this mean? It meant that, without hesitation, we could transfer the knowledge and experience gained from West Coast hatcheries to ours. We were dealing with natural environments that are somewhat unpredictable, but the dimensions and attributes of each hatchery needed to be evaluated and its suitability understood.

Oregon had stipulated and guided us with some commitments. One, we had agreed that we would feed the coho on the Oregon Moist Pellet diet. This diet, for the first time, had produced a first-class, superior fish in survival and subsequent abundance.

Hatchery Needs

We had two obvious problems. One was that, as in most hatcheries at that time, arrangements for storing fish food were satisfactory only for dried items. Oregon Moist Pellet would require refrigeration, and although we had some equipment, it was inadequate. Refrigeration units would have to be added.

Second, unexpected at least for me, was a strict rule that required state agencies to purchase needed supplies and materials only from Michigan producers. Exceptions could be made, but only when we could demonstrate that the desired product was not available from a Michigan source.

It took a few weeks, but we solved these problems. We described the salmon program with enthusiasm, and the bureaucracy approved our need to purchase Oregon Moist Pellet from Oregon sources. We also purchased and installed refrigeration in time.

But our hatchery problems were more complex than that. Don't forget that we also had the kokanee, which by this time had hatched from the egg shipment we had received from Colorado in the fall of 1964. We had planned for that, and the kokanee produced from those eggs would be stocked in inland waters in early spring 1965.

Changing Priorities

However, we had a much larger problem in that the Michigan Fish Division had been spending almost one-third of its budget rearing catchable-sized trout for release in Michigan streams. To raise trout to the size of seven to nine inches usually took two years. In 1965, a very large portion of our hatchery space was occupied by trout intended for release in the summer of 1965, and those intended for release in 1966 were already in our hatcheries, probably in the two- to four-inch size. Remember that our planned release of the coho called for rearing those fish to a size of six or seven inches. We needed that space, currently occupied by the trout, for the coho trial program. Thus, we changed the mission of our entire hatchery system, closing the program that had planted catchable-sized trout, to fit the salmon program.[1]

Several years of research by the department had shown that releasing catchable-sized trout into our streams was a waste of money. A very large percentage of Michigan trout streams provided very adequate natural reproduction of the existing trout species. The hatchery trout were shown to survive only briefly and to provide very limited additional fishing for anglers. It simply was not a wise or a justifiable form of conservation, and it was expensive.[2] By closing this program we would make available nearly all the hatchery space that the state possessed at that time.

Consider that our inland lakes and streams constitute about 3 percent of the surface area of the water within our boundaries. Therefore, when we shifted our course to include the management of the Great Lakes, we undertook a gargantuan leap in responsibility. What we were proposing—the stocking and management of the upper three Great Lakes—would, we predicted, require a very, very large increase in hatchery capacity. We would also have to greatly improve our efficiency.

Reviewing the Situation

We asked for all kinds of information on our existing hatcheries from the hatcheries staff and from field biologists where the hatcheries were located. In this process, several things became apparent, one being that the total capacity of our hatcheries would not produce the number of smolts needed to adequately stock a single Great Lake—Lake Michigan, for example. When we included Lake Superior and Lake Huron, which was proper for us to do, we had a tiny fraction of the needed hatchery capacity.

Something else became apparent that was painful, but that we had to deal with. Some of the hatcheries had been functioning to raise rainbow trout and sometimes brown trout from the egg stage up to only a very, very small size before transferring to a different hatchery. These hatcheries had some limiting factor, usually in the volume of water available. The reports revealed to us that some hatcheries in Michigan never had been efficient and probably had been built at those locations, on those water supplies, decades before fisheries emerged as a science. I suspect that some were authorized based primarily on political judgments.

During 1965 and in early 1966, Wayne and I toured most of the hatcheries and reviewed the situation to confirm what we understood from paper reports and to decide whether to continue or to close them. Bottom line, some of these smaller hatcheries would have to be closed to achieve maximum efficiency and budgetary frugality. Many employees in those areas had worked for nearly a full career, and the closing of their hatchery would impose serious hardship. So, we did our best to ameliorate the situation, offering other job opportunities wherever we could. The process was not without pain, but it was necessary, and we did it.

The one that I thought was most painful is still vivid in my memory. One of the most remote, isolated Michigan communities is in the extreme western end of the Upper Peninsula, close to the Wisconsin border. A small hatchery had existed for decades near the little town of Watersmeet.[3] We went to the hatchery to meet with the staff and to observe the physical elements of that station. We found a beautifully maintained set of buildings and grounds with exquisite handmade cabinets throughout. We met with the staff of three or four people. They saw us coming and were apprehensive. Only a very, very small amount of water was available there—simply not enough of a water resource for a fish hatchery. We decided to close it and made every effort to find employment for the staff members in other hatcheries. However, I doubt that anything we had to offer was acceptable for those people with a lifetime in that remote small-community atmosphere.

Streamside

Another question loomed. In our commitment to the State of Oregon, we promised to release the younger salmon from streamside locations. We could expect them to return to facilities where the spawning adult fish could be captured and the next generation of eggs collected.

We settled on the rearing station on the Platte River near the town of Honor as our first location.[4] From this unit, we could release the young salmon into the Platte River, and when they achieved maturity, their homing instincts would bring them back to this location for egg-taking.

Our second choice—at the headwaters of Bear Creek—was not far away. It was the only sizable tributary to the Big Manistee River, where we anticipated another group of salmon returning. At that time, the Big Manistee, downstream from Tippy Dam, was subject to the extreme fluctuations of water released for hydropower and judged by us, with the exception of its Bear Creek tributary, to be unsuitable for natural reproduction of salmon.

Then we went to visit the Platte River rearing station where Lyle Newton was superintendent. Of course, he had heard that other hatcheries and rearing stations

Platte River State Fish Hatchery c. 1965
MICHIGAN DEPARTMENT OF CONSERVATION/NATURAL RESOURCES

had been closed. When he received word that we were coming to review the facilities under his supervision, I'm sure he expected that we would close his unit, too.

With Superintendent Newton, we toured the very limited facilities that consisted, basically, of some streamside raceways with gravel bottoms. Upon completing our site review, we informed Lyle of our intent to make this the release site for the first salmon introduction. His joy and relief in knowing that he and his job would not be terminated was lovely to behold.

Preparing
to Launch

Financial Challenges

I n those early days, we were desperate to obtain the money that might meet the predictable needs of the salmon program. Dr. Tody was always an expert in looking ahead and projecting our costs for needed equipment and people. Very early he warned me that we would not have enough money to feed the salmon. The Oregon Moist Pellet was an expensive diet, and we had not budgeted for these costs.

Assistance from Above

Then someone on staff brought our attention to some federal legislation that potentially held the answer to our financial problems. The suggestion was that we look at the "Anadromous Fish Act."[1] This act provided some special funding to fishery agencies in ocean coastal states to support the "research and management of anadromous fishing programs." It had been written several years previously, chiefly to benefit the states of the West Coast and Alaska.

The word "anadromous" in the statute specifically states that these species reside in salt water and enter fresh water only to spawn. Numerous species of saltwater fishes follow this pattern in their life cycle—salmon, some trout, alewives, sea lampreys, smelt, striped bass, and shad, to name just a few.

Our salmon program would not qualify, though, since our salmon did not go through a saltwater phase and thus were not, by that definition, anadromous. To restate the obvious: our Great Lakes are not salt water.

Could we possibly get this act amended so that our salmon program would qualify because their life cycle was so similar? We decided to try. Several levels of approval were required. Could we get through the process in time? Again, Director MacMullan's agreement was relatively easy, and it was with his help that we got the necessary additional approvals required, principally from the governor's office. Anytime state officials seek to change a federal law, it must be approved by the governor. With Governor Romney's consent, we sought help from our congressional delegation. We asked if the legislation providing funding could be amended to include fishes in the Great Lakes that returned to streams to spawn. In other words, fish with the same life-cycle pattern, but coming from the fresh waters of the Great Lakes, should also be considered anadromous.

Again, the right person was in the right place at the right time. The name John Dingell had been prominent in the U.S. House of Representatives for two generations. John Dingell Sr. was one of the authors of the Dingell-Johnson Act, which provided for the distribution to states of funds collected as part of an excise tax on fishing tackle and related items. He had been succeeded by his son John Jr., who was always prominent in matters related to conservation in his native state of Michigan. We asked if he would help amend the law, and indeed with his leadership, it was amended, and some funds became available to us in time to be useful. I marveled at how we could be so fortunate, making our salmon program financially secure for the near future.

Legislative Support

While the funds from our new eligibility within the federal statute were very helpful, they, too, would ultimately prove to be inadequate. The director and the commission authorized the Fish Division to seek another special appropriation of a million dollars from the state legislature. To the best of my recollection, it moved through the House of Representatives quite expeditiously, although somewhere along the line, it was reduced from $1 million to $500,000.

Keep in mind that we had yet to stock our first salmon. The support and any excitement were based on our publications and our speeches promising good things. Nothing else had happened—yet.

Then one day I got a call from Senator Joe Mack of the Upper Peninsula (UP). He was a frequent opponent of matters relating to the conservation department, and he was also chair of the Senate Appropriations Committee. Senator Mack requested that I meet with him in his office. He asked a few questions about the salmon program, then asked where we intended to release the fish. I explained that the best biological fit was Lake Michigan, and that the fish would be released in the Platte River and in a tributary to the Manistee River.

Senator Mack was always direct. He said words to the effect, "Dr. Tanner, if you want that $500,000, some of those fish must be released in my district," which included counties in the western UP, bordering Lake Superior. In response, I'm sure I said that I would need to talk to Director MacMullan.

I recall very vividly my conversation with the director—he was quite profane. Our discussion centered on two things: one, we needed the money, and streams tributary to Lake Superior would provide for the needs of our salmon. Lake Superior, in our judgment, would benefit by the stocking of salmon, but it simply wasn't as high a priority as Lake Michigan, with its much greater abundance of alewife and a much greater public demand for recreation.

Long story short, however, we decided to agree with Senator Mack, and about a month afterward, the first salmon were released in Lake Michigan tributaries. About 225,000 smolts were also released in the Big Huron River, a tributary to Lake Superior and in Senator Mack's district. This planting of salmon would provide spawning runs in the years ahead and would become a part of the salmon population in Lake Superior.[2]

Making the Case

There were the most exciting, challenging, difficult, and rewarding months of my life. A logical starting point is early 1965. By late spring, things were going well, and the kokanee was planted with a great deal of fanfare.

Cutting Out Catchable Trout

I announced that the catchable trout program would end when the fish already in the hatcheries were released that summer. Many people predicted that there would be a huge outcry from the public when we ended that program and that a multitude of angry anglers would be demanding my dismissal. Therefore, many of my weekends and sometimes evenings were spent giving many talks to community groups—particularly sportsmen's groups—explaining why it made good sense to close the program of stocking catchable-sized trout.

I used a ploy, usually at the end of each presentation, intended to deride any importance of stocking catchable-sized trout. I would say something like this: "I have heard rumors that some folks oppose stopping the stocking of catchable-sized trout during the trout season, but I think that I know trout fishermen pretty well.

I don't think a bona fide trout fisherman has any interest in chasing that hatchery truck to find out where those dumb hatchery fish are going to be stocked so that he can catch his limit of trout. Now if anyone in the audience considers themselves to be a bona fide trout fisherman and is seriously concerned about when the last hatchery truck has been here and when the next hatchery truck is coming, quickly stand up and tell me!" Obviously, no one ever responded to that.

I received about six or eight angry letters. Who were they from? They were from motel and gas-station owners saying that the closing of the catchable trout program would hurt their business. Bottom line, stocking hatchery trout is not good conservation. It was an expensive effort to provide brief and artificial opportunities to catch a trout. That is not what good fish management is all about.

There were many challenges ahead, but if we had challenges, we also had many assets.

Young Coho

The young coho salmon were doing well. A great percentage of the eggs had hatched, and they were receiving tender loving care by our most experienced fish culturists. Most of the young salmon were in the hatchery at Oden. We had managed to obtain some additional funding and could ask for more in the next budget year.

As luck would have it, the annual meeting of the American Fisheries Society was in Portland, Oregon, that year, and we made plans to attend. Dr. Tody also used that opportunity to meet and make good connections with experienced fishery biologists/culturists in Oregon and Washington. We began recruiting some new staff members, too.

Promoting Possibilities

We needed to sustain interest and excitement, so we made every effort we could in that direction, including producing three articles and a report in 1965–66. That year is important because it was after we received the first eggs, but before we had any clue as to the magnitude of our success. The publications represent our best efforts to predict how and what would be needed to establish the salmon fishery in the Great Lakes.

Dr. Tody and I wrote a couple of articles for the department's magazine—*Michigan Conservation*—about our expectations for the Great Lakes fishery that we were trying to build. "Three Fine Fish" appeared in the March-April issue and described the kokanee salmon and coho salmon options that we were implementing, as well as the striped bass that we were considering. Remember that we obtained kokanee eggs first, and we would be releasing those in two of our large inland lakes—Torch and Higgins.

Striped Bass

The third fish was the striped bass, which we seriously considered introducing. We collected some two hundred of that species from a South Carolina hatchery and transported them to the hatchery at Wolf Lake, where they were held for several months. Without experience and with little definition to our early plans, we had considered that perhaps the southern portions of Lakes Michigan and Huron, and possibly Lake Erie, would be warmer and suitable for first-rate bass rather than a trout and salmon mixture. The more we looked at the striped bass, the more doubtful we became.

We were well aware of the enormous potential changes that, if successful, our establishment of new species in the Great Lakes would involve—biologically, socially, and economically. We gradually reached the conclusion, the more we thought about the cold, moderately productive waters of the Great Lakes, that the striped bass would not be a good fit. Its reputation had been developed in waters substantially warmer, in a climate with warmer winters, and within a public somewhat accustomed to striped bass as an anadromous fish.

We disposed of the fish at Wolf Lake and dropped all plans to introduce striped bass. We had, to some extent, built up the public's expectation for striped bass and we, especially I, received extended criticism from some outdoor writers for this decision. However, with the ability to look back more than fifty years, I believe the decision was the correct one. Your judgment is open to challenge anytime that you introduce a new species into new environments, especially the enormous environments of the Great Lakes. You want to be very, very sure that your choices for introduction are biologically and culturally correct.

I concluded the "Three Fish" article thusly: "Follow this story of these three fishes . . . We'll have some successes, some failures, but hopefully, we'll end up with greatly improved sport fishing for our Great Lakes." Looking back, it seems that that statement was truly prophetic. Ultimately, we decided against introducing the striped bass; the kokanee did not succeed in our inland lakes; and the coho was going to be an outstanding success in the Great Lakes.[1]

"Great Lakes, Sport Fishing Frontier" was published in the November-December 1965 issue, with this appeal for public support: "The Great Lakes can be the greatest freshwater fishery in the world, but they certainly aren't at the present time. Public understanding and support are needed to translate planning and dreams into intelligent action-programs . . . While we can continue to improve fishing in our inland lakes and streams, the sport fishing frontier of Michigan, and indeed of all North America, is on the Great Lakes."[2]

I also projected a critically challenging scenario: "The end of the long, rigorous campaign to control the sea lamprey and re-establish the lake trout is now foreseeable. But the victory may well be lost to the alewife before it's won from the lamprey . . . If we can place a predator on the alewife that will be of interest to sport fishermen rather than concede the low value alewife harvest to commercial trawlers, we can promote sport fishing as well as help to solve the alewife question for commercial purposes."

The report to the legislature—*Coho Salmon for the Great Lakes*—was more technical and not formally published, but was produced by the Fish Division in late 1965–early 1966. Wayne Tody did most of the writing with assistance from me, so we shared authorship, with Wayne listed as senior author.[3]

Conservative Expectations

Some of our expectations were spelled out in the articles and report, and it was clear that we were following the standard admonition: be conservative. We described what we believed reasonable to expect, and spelled out a plan that would take several years to begin to produce enough salmon returning to the streams to make Michigan independent of West Coast sources for salmon eggs. We described plans to have Oregon, Washington, and, later, Alaska provide us with a sustained source of eggs for as long as necessary. I also gave numerous speeches to any fishing group that I could reach, explaining why Pacific salmon were going to make an exciting fishery.

A single smolt ready for release into Lake Michigan.
MICHIGAN DEPARTMENT OF CONSERVATION/NATURAL RESOURCES

When you are introducing fish, in this case coho salmon, into a vast environ-ment, you would customarily expect a return of less than 5 percent; that is, for every one hundred young smolts released, you might expect to get five returning as mature fish to the spawning stream. We used that as our plan to predict how many eggs we would be able to collect as we began the sustained effort in the years ahead.

Wayne spoke of fish averaging five pounds; I occasionally used the figure six to seven pounds. We simply had nothing in our background that would lead us to the size and numbers that were almost instantaneously achieved in the mature return of 1967.

In late December 1965, as promised, Oregon provided us another million coho eggs, and again we were faced with limited hatchery space and the need to buy the expensive Oregon Moist Pellet food out of state. Director MacMullan authorized me to go to the legislature for another million-dollar appropriation for the Great Lakes program. Keep in mind that we had yet to plant our first salmon. The support and any excitement were based on our publications and our speeches promising good things. Nothing else had happened—yet.

I've gone into some of the things that were written about our predictions for how the salmon program would develop. With the vast benefit of hindsight, I can chuckle. Advice commonly given to researchers and fish managers is always to speak conservatively. Don't exaggerate benefits and expectations in the early days of any program. As I look at what we predicted, we clearly followed that advice in general.

We were fortunate. We took the actions necessary to achieve success, and we were supported in many ways by many individuals in a similar frame of mind dedicated to taking whatever action was needed to respond effectively.

New Personnel

As you can probably well imagine, I often think back to the early months of the Michigan salmon program. I usually end up marveling at our audacity when we decided to build a Great Lakes sport fishery.

The cold, hard facts were that we were undertaking something that had never been done successfully in the Great Lakes, which was to establish the coho salmon in waters other than their native range of the northern Pacific Ocean. It had been tried many times before in the Great Lakes, and it had always failed. Further, we lacked the money and the hatchery space.

I'm probably somewhat biased, but I think that the Michigan Fish Division staff then would rank near the top of any comparable unit in the country. People were well trained, experienced, highly competent, and dedicated. They were feeling excited. They shared the goals and challenges of, for the first time, managing the Michigan waters of the Great Lakes. However, not one of us—not I, not Wayne Tody, none of us—had any experience with managing sport fishing on the Great Lakes.

We set about the project in several ways. The first was to hire more people. We had some vacancies and money to fill those vacancies, and that would prove to be very helpful. I hoped to get people with fresh ideas and experience outside the existing mentality of focusing on inland fish management. Fortunately, during this time we were able to hire Myrl Keller. Myrl brought with him priceless credentials. He was a third-generation commercial fisherman from the Saginaw Bay area, so he also knew commercial fishermen. He could pick out those who were law-abiding and those who weren't. He had a degree in fisheries from Michigan State. He had also served several years in the United States Navy, including participating in an expedition to Antarctica. When we first contacted him, Myrl was working for the Ohio Department of Natural Resources, but was anxious to return to his home state.

He brought a great deal of knowledge and skill relating to boats and ships that would be required in our Great Lakes program. We had selected the design for our first Great Lakes boat—the SV *Steelhead*. He located a shipyard in Escanaba that would build it for us, and, yes, I bought it illegally with funds from the category known as supplies and services just before my departure from the Fish Division. But it was Myrl who put it into service in our early attempts to regulate commercial fishing activities that were being phased out to accommodate the salmon program.

Myrl designed a sampling and monitoring program for all of Michigan's Great Lakes waters. He understood what we needed in the way of basic data upon which to base our management strategies. He established crew assignments and schedules for the collection of fundamental facts concerning the new fishery as it developed. He first established a Great Lakes station at Charlevoix and subsequently stations at Alpena, Marquette, and on Lake St. Clair. He was instrumental in developing all aspects of Michigan's Great Lakes program. Later he moved from a staff position in Lansing to be chief of the station at Charlevoix. His service was so important to me, to Wayne Tody, to the department, and to Michigan. Myrl is now retired, and I count him as a friend and I wish him very well.

As luck would have it, at the 1965 annual meeting of the American Fisheries Society in Portland, we were able to start recruiting some new staff members. Eventually, Wayne hired Dave Borgeson and Tom Doyle, who joined the department in 1966.[1] Both were born and educated in Michigan, but both also had several years of West Coast experience. They would serve extremely well on the fisheries staff in Lansing. Many times, their outside experience would prove invaluable.

The First
Release

April 2, 1966—the Official Beginning

t was early 1966, and everything was on track; everything was in place. We had done everything we needed to do, and our next step was to prepare for our first release. Judging by the size of the young salmon in the hatcheries, they would be ready for release at six to seven inches long probably in early April.

We had overcome many serious obstacles. The timing of each step had so often been critical. For the first time in at least a decade there had been a surplus of coho on the West Coast. We had been very fortunate that Oregon was willing and able to send us a million eggs. Perhaps my acquaintanceship with biologists of that state had helped gain approval and cooperation. This event came just weeks after my return to Michigan as chief of fisheries. Evidence of effective sea-lamprey control was in place on Lake Superior.

New leadership was at the helm of the Department of Conservation in the persons of director Dr. Ralph McMullan and deputy directors Harris and Shapton. We had broad-based public interest and support, certainly a **critical factor**. With Bill Mullendore's guidance, we had effective communication with the public and with the legislature. With kokanee salmon, we had sparked public interest. Politically, the opposition that came from the commercial fishing community and their supporters in the federal government, represented by the U.S. Bureau of

Commercial Fisheries, had been ineffective. The unusual and very fortuitous success of pink salmon introduced from a hatchery in Canada probably also encouraged optimism for our success.

Very important and most amazing was our success in gaining additional financial support with people in the right places. They had enabled us to amend the federal Anadromous Fish Act and thereby gain desperately needed dollars. Twice our state legislature added funds—a total of $1 million—required for the special Oregon Moist Pellet salmon feed.

Making a Splash

We began to plan the details of the official release event. A speaker's platform was erected on the banks of the Platte River at the location of the small, primitive rearing station near Honor. Newspapers from all over the state were invited. TV crews were present, and we had a golden bucket with the date printed prominently.

We invited Governor Romney, but he was unable to attend. Instead, we selected state representative Arnell Engstrom to release the ceremonial first bucket of coho salmon. Engstrom was chairman of the House Appropriations Committee, and the Platte River was in his district, so he was a logical second choice. Also present on the speaker's platform was Carl Johnson, longtime member and then chairperson of the Conservation Commission. Troy Yoder, the conservation department's Region 2 director, was there. Seated at the back of the platform was a representative of the U.S. Bureau of Commercial Fisheries, who maintained a dour expression throughout our ceremonies. Standing at the podium, I made a few comments describing what we had achieved and our vision of the future for fisheries. My key phrase was that we expected to create the world's greatest freshwater sport fishery.

In an area prepared for our audience on the opposite side of the river were perhaps two hundred people. I recognized many as individuals who had provided significant support for our efforts, and, most importantly, my mother and father. After I had made my thank-you and other comments, Rep. Engstrom proceeded to the river bank, was handed the bucket, and released our first "official" coho salmon.

The crowd dispersed, and the reporters left. A week earlier, our hatchery crews had released several hundred thousand coho into the Platte River and an approximately equal number in the headwaters of nearby Bear Creek.

I was able to stock a smolt before leaving the scene.

MICHIGAN DEPARTMENT OF CONSERVATION/NATURAL RESOURCES

Mixed Emotions

At one of the tiny tributaries of Bear Creek, in a scene arranged by a cameraman, I made my own personal release of coho salmon. I am sure that each of us has had moments in our lives that are truly unforgettable. This is one of mine. After the demands and hubbub of that memorable day had subsided, I was alone. It was a cold spring day, and a few big snowflakes were falling. I watched the young salmon begin their migration downstream. We in the Fish Division had done it. Everyone could be proud.

I was optimistic; I was sure that we were witnessing the beginning of developing a world-class fishery, but I was also sad in this very special moment. I knew that I would soon be leaving the Fish Division for a post with Michigan State University. It was not yet public knowledge. I had made the decision for several very good reasons. It was the right thing for me to do at the right time, but it was not easy.

The Summer of 1966

left the Fish Division in June of 1966 to return to MSU, and I am sure that my departure came as a surprise to everyone in the department.[1] My choice to leave at that particular time was one of the most difficult ones that I ever had to make.

Five academic units within MSU's College of Agriculture made up a group known as the natural resource departments. They were Fisheries and Wildlife; Forestry/Wood Products; Park, Recreation and Tourism Resources; Resource Development; and an unlikely appendage—the School of Packaging. The large, beautiful Natural Resources Building was under construction and nearing completion. It would house four of these departments. The dean of the College of Agriculture—Dr. Tom Cowden—was seeking to give them an administrative head within his staff.

The first I knew of this was when Dean Cowden asked for an appointment and came to talk with me. A series of meetings culminated in his offering me the position to be known as director of natural resources for Michigan State University. MSU was my alma mater, and the pull was strong. The education that I had achieved there was with the help of outstanding faculty. My principal professors—Drs. Ball and Tack—were urging me to accept the position, even though I was happy where I was. I had exciting programs underway, and I was functioning in my own area of expertise as a leader of an expanding fisheries management program with great potential. I declined to commit.

However, I had personal considerations as well. This was the mid-1960s, and the youth of this nation were in turmoil. Protests against the Vietnam War were producing violence and riots on campuses. Helen and I had three teenaged sons at home. We had no serious problems within our household, but the external threats to our stability were real. I was away a lot, with many weekends consumed by meetings and speeches.

The position was financially attractive. The university was offering a full professorship with tenure and a salary of $25,000. Let me put that in perspective. When I left Colorado twenty-two months earlier, after twelve years of working experience, I was making $10,800. At the Fish Division at that time, I was being paid $15,000. Financially and from a security sense, the offer was extremely attractive. Helen and I spent several evenings in discussion. She always made it clear that it would be my decision, but it was also clear to me that, from a family sense, I should accept the offer.

When I still declined to give an answer, President John Hannah asked me to come and see him. President Hannah could be very convincing, and he spoke of his interest in natural resources. Although he never promised a separate school, he clearly indicated his interest and support. At the end of the interview I accepted, we shook hands, and my life took a new course.

A Turning Point

Perhaps I knew it then, but certainly with the hindsight that has come since, I could define that moment as a major turning point. All my life up to that moment had been oriented towards fish and fishery interests. From a boyhood fishing with my father and teenaged fishing guide to World War II experiences, to graduate education, to my early employment in Colorado and, most recently, as chief of Michigan's Fish Division, everything had clearly been within the scope of my personal and professional expertise. In this new post and in all subsequent positions, I would oversee professional people in a variety of disciplines beyond my personal scope of education and experience. That was a new world, bringing its own problems and rewards, but life in those executive roles would always be different.

However, the new professional path I took did not diminish my personal interest in fishing or my professional interest in fisheries management, especially the marvelous developments occurring in the Great Lakes fishery.

New Hatchery Approach

I did feel a twinge of guilt for having started a large program, getting it successfully underway and funded, but then leaving it before the first return of adult fish. In general, I felt confident that my chosen successor, Dr. Tody, would provide the necessary leadership. As I thought more about the overall situation, however, I came to realize that

1. According to the estimates that Wayne and I had made, there was not enough salmon spawning habitat to fully stock the Great Lakes.
2. That meant that we would have to provide substantial stocking from hatcheries, probably for an indefinite period.
3. The Michigan hatcheries in general had been built more than fifty years earlier, and their design, water supply, and location were totally inadequate for the anticipated need to stock the Great Lakes.

One area in training and experience that Dr. Tody lacked was an awareness and/or appreciation of a new science expressed in the huge new fish hatcheries in various western locations, particularly along the Columbia River. Therefore, I scheduled a tour of a variety of hatcheries on the West Coast, taking Wayne with me. The express purpose of this trip was to expose him to what Michigan would have to build in the way of a new hatchery system.

We went first to Colorado, where, with old friends of mine, we toured the two new hatcheries there—one at Salida and the other on Rifle Creek. From there we went by air to Idaho to visit several hatcheries—both state and private. The Idaho experience was an important part of the trip, but the hatcheries there had one huge asset that could not be duplicated. In general, the cluster of fish hatcheries in Idaho are all located on the equivalent of groundwater with near perfect year-round temperature for rearing trout and salmon.[2]

From there it was on to the Columbia River, where on both sides—Washington and Oregon—were numerous state and federal hatcheries of very large productive capability. All were striving to reestablish huge runs of salmon in the Columbia that had been seriously diminished, apparently by the construction of hydroelectric dams.

Wayne and I exchanged excitement and expressions of wonder in those few days of my remaining service with the Fish Division. Two specific results can be attributed to the trip. Wayne renewed and strengthened his acquaintanceship with Dr. Loren Donelson of the University of Washington. Dr. Donelson became the source of a million chinook salmon eggs that Wayne obtained later that year. The second related series of events was in the design and construction of Michigan's greatly expanded fish hatcheries. Wayne achieved exceptions to a general rule and hired engineers from the West to design Michigan's system. The actual construction process was performed by Michigan companies.

The Jack Run

The well-publicized ceremonial release of coho smolts in the Platte River on April 2, 1966, was past, and a long summer of waiting lay ahead. Cooperating biologists from Oregon and Washington had outlined for us what to expect. If things were progressing well, we would have a precocious run of salmon in September and October 1966. If things went according to plan, a portion of our newly introduced salmon would mature a year early and appear at the parent streams—the Platte River and Bear Creek. These precocious males, commonly referred to as "jack" salmon, would be our first clue about how things were going.

Wayne Tody and I had fished the Pacific in late summer of 1965, and we used photos of those fish to begin to try to prepare our fishing public for what to expect. Jack coho salmon returning to the Columbia River might commonly weigh between two and three pounds. The size and numbers of our salmon provided us with our first clue of success or failure and predicted the first run of mature salmon due in the fall of 1967. In April, we had asked that anglers "show their sporting spirit and help future coho fishing by not creeling any fish under the seven-inch limit that covers trout."[1] That proved to be unnecessary when our "jack run" materialized.

A taste of things to come—Tody (*left*) and Tanner (*right*) with their West Coast coho catch.
HOWARD A. TANNER COLLECTION

We have lost something that was present in the news coverage of that time. Newspapers were far more prominent in our delivery system of news than they are today. Each of our major newspapers and some of the minor ones had an outdoor writer—James Crowe of the *Free Press*; Tom Mowbray of the *Detroit News*; Ken Peterson of the *Flint Journal*; Ray Voss of the *Grand Rapids Press*; Gordie Charles, a freelancer who also wrote for the *Traverse City Record-Eagle*; and several others. There were also two outdoor TV shows—the long-established *Michigan Outdoors*, then owned, operated, and presented by Mort Neff; and *Michigan Sportsman*, which Jerry Chiappetta produced. The news media writing outdoor stories in general-news publications was much more extensive than it is today.

Receiving eggs and the initial planting events were well covered, for example by James A. O. Crowe's *Free Press* article "Vision of a Fishing Paradise," but each provided no actual fishing opportunities until the jack salmon run in the fall of 1966. Yes, it was a new fishery, but it was exclusively in the Manistee and Platte Rivers, with familiar territory, familiar techniques, and very interesting fish of mostly four pounds with an occasional larger one, but otherwise not terribly different from the existing steelhead fishery. Tackle was essentially the same: boats, motors, locations, the style of the trip—with or without guides—were all pretty familiar to at least a portion of our fishing public.

For those readers who were not present and who don't remember those early days from personal experience, try to picture the activities, interests, perceptions, and expectations of the people who were fishing Michigan waters prior to the salmon introduction. The largest group of our fishing public went for bluegill and fish of one sort or another from inland lakes. We had trout fishermen from the dedicated fly-fishing group and all the others who fished Michigan trout streams. There were bass tournaments; smallmouth and largemouth bass were considered important. Pike fishermen occasionally caught a fish heavier than ten pounds, but they were so rare that, if they did, they probably were presented on the next outdoor television show. Very few anglers pursued muskies, mostly on Lake St. Clair and rarely in other inland waters. The lake trout populations that had once supported a very limited form of sport fishing on the Great Lakes were so severely depleted as to no longer constitute the basis for an open-water fishery.

Alewives Yet Again

As expected, an alewife die-off began in May 1966 and continued to the last weeks of June or early July. During the preceding three or four summers, more dead alewives appeared than in the previous year and made for troubled beaches. The year 1966 was not an exception; the beaches were a mess. Somewhere along the shore of Lake Michigan, some dead coho were found, and probably fewer than a dozen of these fish showed substantial growth.

Later that summer, however, jack salmon began to appear in the lower Manistee River. Some coho were caught in commercial nets, and a sport angler caught some, too. Then came hundreds and then several thousand salmon in the lower Manistee River, creating some local excitement. Dr. Tody invited Cliff Millenbach and Ernie Jeffries from Oregon to come and view what was happening. By this time, several hundred had been caught. The average size was spectacular. Four- and five-pound fish were common, with a maximum of seven pounds!

It wasn't long before
sport anglers caught on
to coho fishing.
MICHIGAN DEPARTMENT OF CONSERVATION/
NATURAL RESOURCES

When the West Coast biologists reviewed what was happening, they said, "You'd better get ready as you're going to have large numbers of fish return next year, and they will be of exceptional size." Biologists who had been involved in the program took note, but the public seemed to pay little attention.

Wayne Tody arranged for me to join him for a day of fishing for these salmon on the Manistee River. We were accompanied by Jerry Chiappetta and a camera crew. It was a great day for me. We all caught a fish or two, and the camera crew got their pictures. Jerry featured this day of fishing on his weekly TV program. However, I was surprised, and I'm sure that Wayne and his fellow biologists were also surprised, at how little public interest was generated.

On the West Coast, it was rare for female salmon to appear in the "jack run." But in the 1966 jack run on the Manistee River were quite a few female coho. They were numerous enough for the Fish Division to take and fertilize a few eggs. I think they succeeded in collecting about fifty thousand.

The "jack" run at the end
of the first season was
a favorable foretaste,
complete with eggs to
start a new generation.

MICHIGAN DEPARTMENT OF CONSERVATION/
NATURAL RESOURCES

Comprehension

I was excited. Every clue said get ready, we're going to have a great success. Checking occasionally with Wayne and other biologists I knew, they were all excited, too. But little excitement was expressed in the news media or among anglers. With a great deal of hindsight, I can see that these events were so different, so alien to the fishing public of the Midwest, that what they displayed was lack of comprehension.

By mid to late October though, the excitement was over, and everyone settled into the winter of 1966–67. I'm sure that Wayne and his division were busy. They now had both chinook and coho eggs, as well as young fish, in the hatcheries. Oregon had provided another supply of coho eggs, and the State of Washington was also able to ship some. Dr. Loren Donaldson from the University of Washington provided the initial supply of chinook eggs after cultivating them in his laboratory.

Promise
Fulfilled

Alewives Reprise

Separate but parallel to the budding salmon-fishing story was the depressing, financially catastrophic, sad tale and continuing saga of the alewife. By then, billions of alewives made up more than 95 percent of Lake Michigan's fishery biomass. With no predator fish to establish population control, millions of these fish died in enormous masses and washed up on the beautiful beaches of that lake's coastline. They were a severe public nuisance and a serious detriment to beach-oriented tourism—ruining beaches and swimming opportunities. The situation demanded a solution.[1]

The alewife die-off began quite early in 1967. This had become an ugly annual event, but the one that year proved to be of monstrous dimensions—the very worst, biggest, and most extended. Masses of dead alewives were observed in the open waters of Lake Michigan, where decaying carcasses repeatedly plugged the intakes for Chicago's drinking water. Large floating masses of dead alewives, hundreds of acres in size, were visible from the air. When the wind blew, these fish piled up on the beaches, leaving windrows—in some places more than two feet high—far exceeding anything previously experienced. One newspaper writer described the situation as a pile of dead fish a foot high and three hundred miles long.

The stench began, and the flies swarmed. The smell was so repulsive that people

Dead alewives littering Great Lakes beaches caused many problems.
MICHIGAN DEPARTMENT OF CONSERVATION/NATURAL RESOURCES

didn't go to the beaches for swimming or other recreation. The odor in the nearby communities was almost intolerable, and beachfront communities were evacuated. The masses of dead alewives moved farther north as spring and summer advanced. The shoreline tourism that characterized so many Michigan cities along the Lake Michigan shore was paralyzed.[2]

Communities and individuals did everything they could to alleviate the problem. Bigger and better beach sweepers were put to work on the beaches. Even snowplows were employed, along with front-end loaders and any other suitable mechanical equipment. Tons of dead fish were transported to landfills.

The Salmon Solution?

With this backdrop of a visible problem and unsuccessful efforts to solve it, a big new approach—to introduce an attractive sport fish that would thrive on a diet of alewives—had made sense to many in the public. When we put forth the logic for introducing Pacific salmon, the abundance of the alewife as a food supply was clearly visible and understood by anyone who cared to look. In the midst of the siege, however, it was hard to remember that the dreaded alewives really represented an opportunity for a great sport fishery.

Finally, by mid-July, the crisis passed and normalcy returned. The cleanup had rescued a significant portion of the summer season, but it was a cause for alarm. People envisioned that this was typical of what lay ahead. It had never been this bad before, and as we shall soon see, it would never be this bad again. But communities and individuals began to anticipate a similar problem in 1968 and to prepare as best they could to deal with it.

We professionally trained fishery biologists and limnologists understood what had probably happened. It was the oft-repeated pattern of an introduced exotic. Having left the restrictions of its home range, an exotic species in its new environment will expand until it exceeds the carrying capacity of its new home, or until other factors such as predation limit its abundance. Quite often a population collapses from starvation, such as what we witnessed with the alewife in Lake Michigan in the spring and summer of 1967. But to the public, it mostly appeared as an introduction to a bleak future summer.

The abundance of this fish that we considered a food supply for salmon was not viewed the same way by many others. Everybody demanded a solution. In the minds of the public, the salmon program was still a promise, but only a promise, for the future. Typically, and understandably, they wanted something done now. However, the abundance of salmon in Lake Michigan displayed by the catch and harvest of tens of thousands of them that fall became the explanation when, in the summer of 1968, we saw, essentially, no dead alewives. The public almost automatically concluded that the salmon had eaten the alewives.

While we as professional fisheries biologists knew that this was not the case— there were not yet enough salmon to make a significant impact on the abundance of alewives—still it was easy for us to argue only briefly that it was not so and then to accept the credit.

Public Support

I don't want to overemphasize the added dimension of public and political support that came to us as a result of this very fortuitous and timely enormous alewife die-off. The support of the excited fishing public experiencing the abundance of salmon for the first time was probably sufficient. However, millions of people who had little or no interest in fishing per se welcomed what was the apparent solution to the annually occurring stench of dead alewives at the shoreline cities and beaches.

Approval and support for continuing the introduction of salmon and the building of an extremely valuable sport fishery was virtually unanimous and a **critical factor** in the salmon success story.

Commercial Fishing Solution

Unfortunately, the solution repeatedly being offered by the Great Lakes Fishery Commission and its action arm—the U.S. Bureau of Commercial Fisheries represented by our longtime adversary Dr. Fenton Carbine—was vividly inadequate and unsuccessful. Wisconsin commercial fishermen harvested alewives and had them processed into fish meal and fish oil. However, this approach quickly failed. The market value of the alewives caught proved to be far less than the cost of catching and processing.

Nevertheless, Carbine and his associates continued to promote plans for an even bigger alewife trawler fishery and demanded that processing plants for producing fish meal and fish oil be licensed in Michigan. Operators of large trawlers (the most efficient way to harvest alewives) petitioned the department to be allowed to harvest the huge number of alewives in Lake Michigan. Dr. Tody, backed by Director MacMullan, refused them a license.

When Dr. MacMullan refused, Dr. Carbine threatened legal action, and this proved to be his professional undoing. This time he had pushed their patience too far, and they took further measures to frustrate any future actions he might take. The office of Michigan's governor officially requested that the U.S. Bureau of Commercial Fisheries' functions not related to research be removed from Michigan, along with Dr. Carbine. The request was granted fairly quickly.

All the bureau's programs that supported commercial fisheries were closed, and over the next few months, twenty-eight members of the Ann Arbor staff were transferred to other stations. Only the very competent and respected research staff remained in the U.S. Fish and Wildlife Service offices in Ann Arbor. This ended any visible opposition from the federal government.

The Dream Come True
and a Nightmare

ll of us who had been a part of the preparations for a Great Lakes salmon fishery were waiting with great expectations. An exceptional run of precocious jack salmon had happened in the fall of '66. The biologists from Oregon and Washington had predicted that there would be an outstanding run of mature salmon in late summer and fall of 1967.

The news media had covered the salmon story well, but for the most part, the sport fishing public in Michigan seemed not to comprehend, or perhaps took a wait-and-see attitude. We saw little evidence of growing excitement. Catching salmon? I suspect that most avid anglers in Michigan at the time had dreamed of a long trip to Alaska or perhaps British Columbia or, to a lesser extent, the diminished salmon fishery of Oregon and Washington. It was something that they might do sometime if they had the time and could afford it, but it was actually unattainable for most people.

Commercial Catch

But some other things were happening. As the ice left in the spring of 1967, commercial fishermen from Indiana began to catch and market coho salmon, averaging four to five pounds each. The urgent attention of Dr. Tody and Director MacMullan produced reasonably prompt action by Indiana's Natural Resources Department to close that fishery.[1] However, before the commercial fishing operations could be stopped, they had harvested an estimated 20,000 four-pound salmon.[2] This was unfortunate, but it was also a wake-up call.

It meant that the salmon were thriving and had moved to southern Lake Michigan. On the other hand, it was also a concern. How significant was this catch? Perhaps they had taken most of the surviving salmon. More wait-and-see during that summer. Occasional dead salmon were found along the shores, but those numbers were insignificant. The fish were being found farther north along the western shore of Michigan's Lower Peninsula and, in each instance, the fish were demonstrating a very rapid growth pattern.

Of course, through the late spring and most of the summer, many folks were also preoccupied with the devastating alewife die-off and trying to rid their lakeshore communities of that scourge.

Following the Salmon

The department adapted existing watercraft to survey the salmon. The sampling began showing good numbers of large coho in the vicinity of Manistee and then northward to Pentwater and Frankfort. When these fish were taken ashore and displayed, a spark of excitement began to develop. I am struggling as I write these lines more than fifty years since that unprecedented, never before seen or equaled event, that burst of sudden availability of the adult coho approaching the bowels of the streams where they had been released. Efforts were made to alert the public, but it seemed that very few paid any attention.

The Surprising Surge

However, in early August, adult coho salmon, much bigger than our original expectations (even with the highly successful jack run the previous year), began congregating off the Manistee River and the waters northward to the point where the river enters Lake Michigan. A scattering of reports began to come in. Venturesome fishermen caught several salmon—one weighing fifteen pounds, a few twenty to twenty-five pounds, and, not often but occasionally, even a thirty-pound salmon.[3] Holy mackerel! Uh, holy salmon!

The news spread like wildfire by word of mouth. The informal network of fishermen, the excited phone calls from friends and family members accomplished what newspaper headlines and TV shows had not. The calls probably sounded something like this as Joe was calling his brother in Detroit. "You won't believe this; I was out yesterday with my neighbor Carl, and right offshore from Manistee we caught seven salmon! They're huge! Our biggest weighed sixteen pounds! You've got to get up here." Sam, from the little town of Empire, was calling his son in Grand Rapids. "Jim, Platte Bay is full of salmon. They're everywhere! They're enormous! They're leaping out of the water! We caught two. Our lines broke three times. They're over fifteen pounds! They're right on the surface! Spread the word and get up here!"

The trickle of calls during the first few days was followed by an avalanche of people descending upon the small lakeshore communities from Manistee northwards—Frankfort, Beulah, Honor, to the launch site at the mouth of the Platte River and on to the small town of Empire five or six miles north. This had never occurred before, and the congestion, fortunately, never since. The Great Lakes region, perhaps even the nation or the world, had never seen a similar event. How could anyone have prepared for it?

As I reflect through hindsight, I realize that, as I had looked ahead, I had thought only as a fishery biologist. I thought about survival and growth and the suitability of the alewife as a food supply. I thought about places for the fish to spawn. I hoped for good survival and good growth and worried about temperature and all the things related to the salmon themselves. I freely admit that I had given little thought to the fishermen and the challenges and implications of having a large-scale unorganized fishing effort on the huge waters of Lake Michigan. Imagine my absolute delight that the fishermen were happy. That was what Wayne and all the others who had worked and waited for something like this to happen had wanted.

Just Imagine

Try to imagine the impact. Anyone in Michigan who valued fishing as a recreational experience wanted to be there—had to be there. None of these fishermen had ever really imagined a salmon fishery in Lake Michigan. Not one in perhaps a thousand had ever traveled a great distance—to the West Coast or to Alaska—and experienced catching a salmon. Now, suddenly, all they had to do was get a boat, a couple of friends, and in a few hours be in the middle of the action.

The full impact of this news had spread in time for many thousands of people to arrive for the Labor Day weekend of 1967. By then, hundreds of boats had come, and that number grew into several thousand. The Fish Division estimated that five to six thousand salmon were caught during the three days of Labor Day weekend. The influx absolutely swamped the service facilities of that area.

The characteristics of the tourism industry in the northern Lower Peninsula of Michigan didn't fit. According to the existing pattern, the tourist season was essentially over following the Labor Day weekend. Families were gone so that their children could return to school. Perhaps more importantly, the workforce that serviced the tourist industry also disappeared: schoolteachers returned to their jobs, college students were back on campus, and other people who had a plan that included summer work had gone back to their other ways of making a living.

I was not there then, but I did make it a few days later, and I had at least three trips on Platte Bay with friends and family during that wonderful, exciting burst of the new Great Lakes salmon fishery.

A Typical Trip

I think perhaps my trips were fairly typical. One of my friends had a fourteen-foot boat on Platte Lake. It had a 5 horsepower (hp) motor; we had miscellaneous rods and reels that we had used to fish on inland waters.

First, we discovered a long waiting line for the very limited launch facilities at the mouth of the Platte River. Hundreds of cars with boats on trailers lined up every morning, often extending back as far as Highway 22. People would wait an hour and a half or two hours for their turn to launch their boat. Launching the boat had its difficulties; the platform was small and inadequate, and we were launching into the current of the Platte River. Complications of one sort or another developed.

Oh, okay, we got the boat launched. Then we had to park the vehicle and trailer. The first available space was alongside the road more than a mile and a half away. Then the driver had to find a way to get back to where the rest of us were waiting at the mouth of the river. Soon enough an entrepreneur appeared out of nowhere, and for two dollars, he functioned as a taxi.

Going coho fishing? First find a place to park!
(MICHIGAN DEPARTMENT OF CONSERVATION/NATURAL RESOURCES)

Then we made our way down the short seventy-five yards of river to Platte Bay. On both sides of the river—standing in waters nearly to their waist, shrouded by their waders, nearly shoulder to shoulder—were lines of fishermen extending northward and southward for what appeared to be a quarter of a mile. So, you have ten thousand or even one thousand fishermen in this small area. How can they fish safely? Where can you find a bathroom? And if you find one, how long is the line?

We scanned the calm waters of Platte Bay, and boats were bobbing as far as we could see. We made our way out into the open waters and saw boats of every description. The typical one was like ours; some were even smaller. We saw two

Fishing boats in a morning rush to leave Manistee for coho country in Lake Michigan.
MICHIGAN DEPARTMENT OF CONSERVATION/NATURAL RESOURCES

canoes forming a makeshift catamaran. Two, perhaps three, rubber boats were propelled by people with oars. I doubt if more than half a dozen boats were longer than ours. Over time, small boats were replaced gradually by bigger boats. In 1968, I borrowed a sixteen-foot long Lund with two 18 horsepower motors on the back.

Every person that we saw was excited and thrilled to be there. The hundreds of boats were filled with fishermen who had never been on the big lake before. Their tackle was inadequate. Their rods were too small and light; their lines broke repeatedly. Their boat motors were not powerful enough, but fish were everywhere. They were hooking fish; they were landing a lot of huge salmon. People who had never caught a fish larger than a four-pound bass had the experience of catching several, maybe five, very large salmon, perhaps all of them fifteen pounds or larger. They might be in a boat with two or three friends and each of them caught that many. They had shared the excitement with many other boats nearby.

One other feature of the salmon fishery then was the coho's habit of porpoising.

Anglers fishing for coho even from a small boat on the big lake could haul in some lunkers.

MICHIGAN DEPARTMENT OF CONSERVATION/NATURAL RESOURCES

People were out there full of expectations when suddenly, close to the side of their boat, ten, fifteen, or even twenty huge salmon would break the water, dive back in, and disappear as if to deliberately tantalize the inexperienced angler.

Seemingly everybody wanted to participate. Everybody wanted to learn and see at the same time. Most of the fishery that year occurred between the port of Manistee on the south and the little town of Empire immediately north of the bay.

Making Adjustments

None of our lures was just right; the available ones were made for catching bass and pike. One of the early popular lures was the flatfish, particularly the silver flatfish. If you know or can remember the flatfish, it had a big wobble, but the hooks were

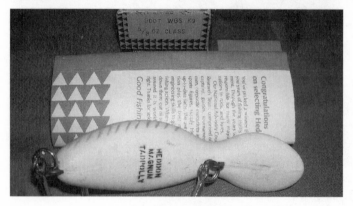

TadPolly lure

very small—totally inadequate for a fish this size, but they worked. Soon no silver flatfish could be had. Other flatfish were painted silver; that supply was exhausted, so flatfish were rented. Then Heddon in Kalamazoo began to manufacture the TadPolly in different colors. My friend Bill Gautsche said their production reached four thousand a day, and they couldn't keep up with the demand.[4]

Keep in mind all the things that we as fishermen did not have and that we now take for granted. We didn't have downriggers—so, Great Lakes fishermen improvised that very useful tool. The first downrigger I ever saw was made of the front wheel of a bicycle. My first downrigger was an eight-inch plank of wood the length of my forearm with a broad notch at each end and the handle mounted towards the end of the plank on opposite sides. Sixty feet of nylon cord, marked with fingernail polish to indicate five- and ten-foot intervals, was my downrigger. Attached to this line was a sash weight—about two pounds of cast-iron with a hole in one end. Back then, windows were raised and lowered by hand and counterbalanced with a sash weight–secured line inside the window frame.

We had no fish finders. There was an occasional Lowrance depth finder—the famous green box. It would give you a pretty good reading of how deep the water was—maybe, just maybe an occasional blip indicating a fish.

I had never thought of carrying a compass on a fishing trip until one morning we encountered fog so dense that we couldn't see fifty feet in any direction. I realized that I didn't know for sure where we were. From that time on we, like everyone else, carried a compass.

It would go on from there in that monumental fall of '67 and in the early years that followed. Nothing was big enough or strong enough. We needed heavier rods, stronger line, bigger landing nets. Adjustments would develop to produce the boats and equipment necessary for safe and effective fishing trips on Lake Michigan.

Where do you put the fish when you get them in the boat? We needed a bigger ice chest and a place to buy ice; we sure couldn't put them overboard on a stringer. The stores in small communities sometimes ran out of everything. Most gas stations for nearly fifty miles ran out of gas. Worse yet, the beer supply in Manistee was quickly exhausted.

You had better bring your own place to stay; hotel accommodations were impossible. There wasn't a room for rent within fifty miles. Fishermen brought their sleeping bags and slept alongside their automobiles, or perhaps in a little house trailer that they had been using for deer hunting. The Moonlight Motel on Manistee Lake in the immediate vicinity of the small boat launch became famous. Some wise person in the Michigan Department of Conservation reserved a few rooms for visiting dignitaries, and it became a gathering hot spot. It was immediately booked solid for fishing for the next two years.

Gearing Up

The demand for new and bigger gear went off the chart. You needed a bigger boat; you needed a bigger motor, maybe two—25 to 50 hp and larger became standard. Larger ice chests, larger landing nets, increasingly sophisticated fish finders, a two-way radio became essential. It was very helpful if you had a friend who was a charter captain and who would signal you where the fishing was the best.

A multitude of smaller needs and functions had to be addressed. Okay, so you catch a salmon. How do you dress a fish that large? Your fillet knife is probably very small. Okay, so you get it dressed. Where do you put the carcass and other awful offal (waste)? Your wife says, "How do I cook this thing?" Yes, there were multiple difficulties and deficiencies, but large fish that were pretty easy to catch made everything all right.

No one had ever experienced such congestion. The excitement, the crowds, the crush of boats during Labor Day weekend of 1967 was totally new to everyone and by any standard. This was a first!

The Nightmare

That period in early September of 1967 was fortuitous in so many ways. For more than three weeks, the waters of Lake Michigan were very calm, which certainly was not always the case. Fall winds and even storms were common. When cold waters are at the surface, persistent winds from the east will produce what is known as an upwelling—pushing the warmer surface waters westward across the lake and bringing a surge of colder water up from the depths to the surface. The salmon had comfortable cold water throughout the top twenty feet.

However, a tragic storm occurred on September 23, 1967.[5] I was alerted to the storm and its effects by a frantic call from my mother on Sunday morning, September 24. She feared that I had been there during the storm and was greatly relieved to hear my voice on the other end of the phone line saying no, I hadn't been there that weekend.

Quickly checking then and listening to accounts and reading accounts since then, these are the salient points of that storm.

Several thousand small boats were fishing for salmon that day. Boats were scattered principally from Frankfort north to Platte Bay, with another entrance point at Empire, seven miles north of the bay. The National Weather Service alerted the Coast Guard quite early in the day to the route and pending nature of a huge thunder and lightning storm coming across and upward on Lake Michigan from southern Wisconsin and Illinois. Its route would strike ports of call close to and including Platte Bay later that morning.

Storm warnings! A red flag on the flagpole was up early, changed quickly to two red flags. A gale was coming. You know that only one fisherman in ten on that day would appreciate the meaning of two red flags. The Coast Guard cutter was out early, approaching the margins of the swarm of small fishing boats. They ordered the fishermen by bullhorn to head for shore immediately. A helicopter flew overhead sounding the same alarm. A small percentage understood the danger and obeyed the commands, but many, probably most, ignored the warnings.

The problem was compounded by the shortage of alternatives. Many had launched at Frankfort, seven or eight miles to the south of the fishing grounds. That was a very long way by 5 hp motor, and that size was almost standard in those days. To enter the mouth of the Platte River to the boat launch facility was a slow process. Only a small percentage of boats could make it in time. Empire was another seven miles to the north, and a boat landing there would have to be made across the beach with no protecting shoreline or piers.

The result was tragedy and chaos. Many boats were swamped; hundreds were landed on the closest beach available, there to be swamped with water and sand from the twelve- to fourteen-foot waves that developed so very rapidly. While hundreds of boats were damaged, the death toll was only seven. When you calculate that probably more than five thousand fishermen were in the boats on the lake that day, we were very fortunate that the death toll was not higher.

Nonetheless, the open-water fishery continued in the vicinity of Platte Bay and nearly a mile from the Manistee River well into October. The word had spread far and wide. The long line of cars and boat trailers along the road to the launch near Platte Bay was a source of information. In the early weeks, they had all been Michigan license plates, but later you could find many from Ohio, Indiana, Illinois, Kentucky, and Tennessee, and even an occasional one from Nebraska or Iowa.

The River Run

Once the adult cohos left the big lake, they crowded into the streams of the Platte and Bear Creek. The story was that the salmon would bite once they were in the river. Some people find the sight of large fish in shallow, clear streams irresistible. They must possess the fish one way or another, and so it was that our problems with snagging and other unsportsmanlike activities began. Of course, the explosion of excitement over the open-lake fishing and even the stream fishing dominated the public's awareness.

Coverage of this developing story exploded in the news media. Not only in the sports section, but on the front page of newspapers and magazines. TV shows were bursting with excitement. The wild experiences and accounts of the return of salmon in the fall of 1967 were news that spread across most of our nation. It truly

When coho salmon
headed "home" to their
native stream, the
fishing folk followed.

MICHIGAN DEPARTMENT OF CONSERVATION/
NATURAL RESOURCES

sparked the excitement and imagination of millions of people. At the end of each year, the news media listed the ten most significant news stories of the year, and for the first time in memory, the *Chicago Tribune* chose fishing—salmon fishing—for the top ten in outdoor news.[6]

The recognition brought important praise to the Department of Conservation, and the Fish Division staff members were heroes. Since 1964, the chiefs had been telling all who would listen that we were going to create the world's greatest freshwater fishery. Now there was clear evidence for everyone to see that our plan was working and what we had promised would really happen.

Swimming Upstream

The open-water fishing and, for practical purposes, the river run of salmon was over for that year. However, "behind the scenes," or rather, upstream on the Platte River, on the Little Manistee, and on Bear Creek—tributary to the Big Manistee River—the "season" was not over. Crews from the Department of Conservation's Fish Division had much more work to do.

Harvest Facilities

Harvest facilities were minimal and primitive, but were developed eventually to manage the surplus harvest. In anticipation of the return of adult coho to the streams of their release, the Fish Division had done its best to prepare. They had hired a contractor to install a weir on the Platte River at the site of the fish hatchery, and also on the Little Manistee River. It might seem a little strange that a weir was being installed on the Little Manistee. After all, the smolt salmon had been released from other streams. However, because such a large number of strays went to the Little Manistee instead of the nearby Big Manistee during the precocious jack run of 1966, it was judged prudent and convenient to begin to build a facility for taking eggs at that location.[1]

Unfortunately, the contractor preparing these installations had not finished as of late October 1967. By then, though, coho salmon had returned by the tens of thousands to be blocked by those weirs. The Platte and the Little Manistee are not large at those locations. They were probably no more than fifty feet wide, and the salmon piled up against them. It was an unbelievable, completely new experience for fishery biologists and the public alike to view those huge numbers of large, beautiful fish.

The Hard Work Continues

My longtime friend John MacGregor—fishery biologist for that district at that time—was directing the program to take eggs from the fish. The department crews there were handicapped, working towards the transfer of approximately fifteen to twenty thousand fish to other streams and to upstream areas of the Manistee River. Because the new facilities were incomplete, nearly all the work had to be done without any mechanical equipment, and nearly every phase of the operation had to be done by hand. Much of the work had to be done standing in waist-deep water, often with snow blowing in their faces, and was truly the hardest of backbreaking labor. It was brutal. The excited fish would sometimes leap from the water, and the crew members would be struck in the chest or head by a large salmon. This was not fun; it was dangerous!

Using seines and other gear, John and his crew spent days netting salmon from the partially completed raceway. They had to capture the fish, sort them into male and female holding areas, take eggs and fertilize them, and load them into trucks to transport properly to a waiting hatchery. Other fish were transported to several experimental upstream areas, and a large number were simply surplus. Some of the surplus fish that were in better condition were sold, some were given away, and many were simply disposed of in pits.

John has told me a vivid story of how he worked twenty-one straight days, averaging 17 to 18 hours each, took two days off, and then did another twenty-three days before the operation was closed and finished in early December. According to his story, he was tired, it was late at night, and his home was in Cadillac, some thirty-five or forty miles from the work site. He said he was speeding, couldn't wait to get home to get cleaned up, get something to eat, and go to bed. He was probably traveling 85 miles an hour. Then he saw the blue lights and heard the siren of a state police car behind him. He braked to a stop, and the officer approached the car.

The officer started to tell him how fast he was going and how he was going to be severely penalized. John said, "Officer, do you know this is a state car?" Unbelieving, the officer went to the rear of the car and wiped off the mud-spattered license plate. About this time, the second police car arrived, and an interesting conversation ensued. When they discovered that John MacGregor was the person in charge of capturing salmon at the weir, their tone changed.

Flashback to a few days previously, when a conservation officer had stopped by in the middle of the salmon harvest and said to John, "What are you going to do with those surplus fish?" John replied, "Take all you want," and that conservation officer loaded his trunk with lots of salmon and went on his way. As it turned out, as a goodwill gesture to his working partners in the State Police, he had stopped at the State Police post and distributed salmon to everyone.

The officers returned to John and said words to the effect that they would let him off with a warning, and to please hold his speed down until he got home. One good turn deserves another.

Assessing Success

After the fishery and its excitement slowed, the Fish Division had some time to assess. They didn't have the resources or a plan to do precise numerical calculations, so the following information is the best estimate that they could make. The mature fish of 1967 were survivors of the approximately 850,000 smolts released in the spring of 1966 (including those in the Upper Peninsula stream in Senator Mack's district). In a general sense, stocking salmon or trout of that size, a fishery manager would be very satisfied with a survival rate of 5 percent.

In this instance, however, with a huge lake full of an abundant food supply and, essentially, no predators, the percentage of their survival and growth had been "something spectacular," far beyond anyone's expectations. Yes, the Indiana commercial fishermen had taken about twenty thousand. I think that Wayne later estimated that the anglers of that late summer and fall had harvested approximately forty thousand fish.[2]

As for the return, the survivors of the early release—now get this—the Michigan Fish Division estimated that they totaled somewhere between 250,000 and 300,000 adult salmon. Big, strong, silvery, and willing to strike—beyond anyone's experience. All told, he and his staff estimated the survival of the original smolts released at perhaps 35 to 40 percent, an unheard-of rate.[3]

With hindsight, such a spectacular survival might have been anticipated. There was a great abundance of food—alewife, anywhere from an inch to nine inches long and readily available throughout Lake Michigan. The young salmon could shift to a fish diet as soon as they entered Lake Michigan. Water temperatures throughout the season were generally optimal and different than most natural environments. Almost no other predaceous fish was available to reduce their numbers.

I have always had a question about the catch that will almost certainly never be answered. In addition to the commercial fishery in the Indiana waters of Lake Michigan during that period, substantial numbers of commercial fishing enterprises were still operating in Illinois, Wisconsin, and Michigan, most of them using gear similar to Indiana's and fishing in a similar manner and timing. I have never heard of any salmon being caught by these other commercial fisheries. Until I hear otherwise, I believe that the crews of those other commercial fishing enterprises simply kept their mouths shut. If I am right, or even partially correct, the survival of that first plant of cohos was substantially higher than even the estimates that we worked with.

Sustaining the Excitement

Another form of excitement emerged. Bill Cullerton, then president of his company in Illinois, spent a month on the scene and projected that it would be the biggest thing the tackle industry had ever encountered. Luhr Jensen of his West Coast manufacturing company spent a month that year and each year following, fishing and choosing designs for new, effective plugs.

VIP Excursions

Trying to expand the interest and support of a large number of people, particularly state legislators, the department began providing a fishery experience for selected dignitaries. Troy Yoder—director of Region 2, which included this new fishery— with the assistance of district fish biologist John MacGregor, began offering a fishing trip out of the port of Manistee on the only sizable boat available to the department at that time: a twenty-two-foot Chris Craft. This boat was not really designed for fishing, but it was the most appropriate one available. These staff members gave many state legislators a first-hand fishing experience.

The news spread, and Illinois governor Otto Kerner and Wisconsin governor

Warren P. Knowles came. Governor Knowles flew his own airplane in on a day when the lake was too rough to fish, so he and John Macgregor flew the coastline from Manistee to Cheboygan. John has recited the story of that trip to me.

The usual Labor Day celebration of the Mackinac Bridge was a focal point. Governor Knowles and Governor Romney walked the bridge together in 1968 and had earlier expressed their desire to fish for salmon.[1] Bill Gautsche was designated to be their host and has described to me how they met in the morning at his home and had breakfast with his family. For two days, they fished together, caught their limits of coho salmon and, of course, were instant converts and supporters.[2]

Other VIPs also wanted some of the salmon experience. Since the summer of 1966, I had been on the faculty of Michigan State University, and MSU president Dr. John Hannah was excited about, and supportive of, the salmon program. He asked me to take him to the Platte River and to show him the run of salmon in the fall of 1968. I took him there, accompanied by Dr. Tom Cowden, dean of the College of Agriculture and my immediate supervisor.

John MacGregor arranged for us to take the trip by canoe. We started at one bridge and traveled perhaps two miles downstream to the next bridge on the Platte River. The salmon were present in unbelievable numbers. Trying to be humorous, I told Dr. Hannah, who was in the bow of the canoe, to make a guess at the number of salmon we saw as we paddled down the stream between the two bridges. It would have been a truly vast number. When we exited the river, President Hannah turned to me and, with a big smile, said "I counted exactly ten thousand!" It was an unbelievable trip, and I thought that ten thousand was as good a guess as any.

Picture Perfect

National press attention continued off and on, with various stories emerging from time to time in publications such as the *Saturday Evening Post*. For example, in the early '70s, I got a call from *National Geographic* telling me that they were planning a section on the Great Lakes for an upcoming issue. They said that they wanted to illustrate it with a picture of me on the lake catching salmon. At that time, the salmon fishery was in full swing, and I readily agreed to help arrange a fishing trip for their photographer. Delays pushed the trip into the month of October, so I told them that the salmon fishery was about over for the season. Still they urged me to try to make it possible, so I collaborated with Traverse City charter captain Rick Zehner. It appeared that our only chance was a small group of coho still surfacing off the mouth of a small stream up the shore towards Suttons Bay. Helen and I met photographer Martin Rogers at a Traverse City motel the night before and listened with interest to some of his world traveling experiences.

The next morning as we prepared to put the boat in the water, it was obvious that the photographer was not adequately dressed for the cold temperature. We had only a short distance to go before we could start fishing, and he sat in the bow of the boat with arms wrapped around his shivering body. He instructed us, "Don't catch a fish yet, there isn't enough light." The rest of us looked at each other, knowing full well that it was going to be very difficult to catch a fish at any time under any circumstances since the season was about over.

He repeated his instruction several times, saying, "Don't catch one yet; there's still not enough light." After another fifteen minutes, he looked up and said, "There's enough light now; go ahead and catch fish." Lady Luck came forth for us. Within ten minutes, we hooked and landed the only salmon of the day, which you'll find on the dust cover of this book. He took many pictures and said, "All right, let's get off this cold lake," and the trip was over.[3]

Speaking of Salmon

The excitement of the fishery of 1967 produced a flood of requests for me to appear as a speaker for various groups and clubs around the Great Lakes. I gave speeches to an assembly of various fishing organizations in Duluth, Minnesota, and talked to the Lions Club in Milwaukee. I also gave talks in Toledo and one as far away as Rochester, New York. The one in Rochester was a banquet in a room that I estimated held between four hundred and five hundred people. It seems that the New York Conservation Department was reluctant or a bit slow to adopt the idea of introducing salmon into Lake Ontario. I was sometimes referred to as the Johnny Appleseed of the salmon program.

Chicago was another important area for speaking engagements, including one at the very large Rotary Club. During the winter of 1968–69, I told my usual salmon story at the annual boat and tackle show at the enormous Merchandise Mart. Other speakers were coming and going to the podium that afternoon, and for our convenience, a coffee service and lounge were offered backstage. It was there that I met and talked with a certain Coast Guard officer who described his experience.

It was September 23, 1967, the day of the furious and deadly storm that wreaked so much havoc, swamped hundreds of boats, and killed seven people. He was the commander of the Coast Guard cutter and had been trying to alert fishermen and ordering them off the lake. Many would pay no attention. The day was coming to a close, the waves were monstrous, and he was headed back to his port in Frankfort.

"We were offshore," he said, "perhaps a mile or a mile and a half. It was safer out there as we rounded Point Betsie. Unexpectedly, we came upon a boat with four fishermen. They had lost power, and their small cockpit between the windshield and the stern was nearly full of water. There were four very frightened, very cold, and very wet fishermen. With considerable difficulty, we got a line to the boat and lifted the four men onboard our Coast Guard cutter. I sent them below deck and took their boat in tow. They took off their wet clothes, were given blankets and coffee, and made comfortable.

"I was extremely busy, and perhaps twenty minutes had passed when my boatswain came to me and said, 'Sir, we're towing that small boat, but unless somebody steers, it is just impossible in this very rough water.' I went below and asked who owned the boat, and one man raised his hand. I told him, 'Here's the situation. We're either going to have to cut that boat loose or somebody is going to have to go back to it and steer it.' He said he would steer the boat.

"We outfitted him with some dry clothes and some rain gear and again, with considerable difficulty, transferred him back to the small boat in tow. It seemed to be working well. I was busy and it was perhaps twenty minutes before I had time to check on him again. I looked back and you won't believe this—there he was, steering with one hand and trolling with the other!"

One morning in Chicago I got up early and took a walk along the lakeshore, including one of the piers. As I was walking towards the end of it, I met a man who was leaving. He was dressed in a business suit. He had a briefcase in one hand and a fishing rod in the other, with a large salmon on a stringer. I stopped to admire his fish. He was beaming and said, "Look at me. I live in Chicago. I work in Chicago, and this morning I caught a salmon on my way to work!" That statement says it all. This fishery is on the doorstep of millions.

Celebrating Success

In early 1968, Wayne organized a large event to celebrate the success of the salmon fishery. It attracted hundreds of people to Detroit's Cobo Hall, where Pacific Coast states were recognized and thanked for their contribution. I enjoyed observing all the positive connections the new salmon fishery had helped to form at the state and national levels.[4]

Over the years, recognition for the successful coho salmon introduction has come in various forms, including numerous awards and honors. Citations and plaques came from conservation organizations such as the National Wildlife Federation, Michigan United Conservation Clubs, Great Lakes Sport Fishing Council, and Trout Unlimited; professional societies such as the American Fisheries Society and the Sport Fishing Institute; and businesses such as Heddon Lures.

In May 2016, Michigan DNR celebrated fifty years of the Great Lakes salmon fishery with a ceremony at the initial release site on the Platte River. Helen and I and our son Hugh were there. I shared some history with the group, then ceremoniously planted some coho salmon smolts in front of the crowd and cameras. Later that month, Helen and I participated in launching and christening Michigan's newest Great Lakes research vessel—the R/V *Tanner*. To put such a vessel into useful service on Michigan waters of the Great Lakes is a truly appreciated memorial.

Further
Developments

Stewardship
and Contaminants

Another very important development that was stimulated by the world-class sport fishery was increased stewardship, the creation and growth of a constituency on behalf of the lakes.

Prior to 1967 and the opening excitement of the sport fishery, the Great Lakes were always just "there." We viewed them with awe and admiration, but seldom did many of us venture out onto them and begin to grasp the full significance of this enormous freshwater complex, the largest such system in the world.

Many more people are now familiar with these magnificent bodies of water. We citizens of Michigan's two beautiful peninsulas are, by geography alone, the stewards, the caretakers of these great freshwater "inland seas." At least partly because of an awakened understanding of the lakes, created by acquaintance with and participation in the fishery, we have stepped forward in any number of ways to preserve the health, the beauty, and usefulness of these waters. Stewardship takes expression in many forms.

Cleanup Concerns

We citizens of Michigan, supported by other states and Canada, have led in developing new and improved regulations to protect Great Lakes water quality. Fifty percent of our people receive their drinking water from the Great Lakes, and we have spearheaded higher standards of water treatment and the cleanup of toxic waste sites. Michigan led in banning DDT and PCBs. I won't go so far as to claim that the Great Lakes fishery created all this awareness, but it is certainly due a share of the credit.

Beginning with the opening of the exciting sport fishing, the period of the '60s and '70s were also the time that a true sense of environmental wisdom developed. Michigan can be proud to have been, in many instances, a leader in this environmental movement. I'll use the state's legislation to properly control persistent pesticides as one example.

The Banning of DDT

I was busy at MSU, involved in administrative duties as director of natural resources for the university, while the effects of environmental contamination by DDT and other persistent pesticides were dawning on us. Rachel Carson's famous and decisive book—*Silent Spring*—had been published in 1962, and its impact was being fully felt. The title of the world-famous book came from studies made on the MSU campus by its renowned ornithologist Dr. George Wallace. It was my pleasure to know him. In his studies, he demonstrated that spraying DDT (Dichlorodiphenyltrichloroethane) on all the trees on MSU's campus was producing a population of night crawlers and earthworms with heavy concentrations of the toxic substance. When robins consumed these prey items, the birds became sick, failed to reproduce, stopped singing, and frequently died. Thus, the robins were silent, so we had a "silent spring."

Wallace and his work were shamelessly attacked, harassed, and vilified by the agricultural community. Many of these leaders, I'm sorry to say, were from the College of Agriculture at MSU. However, Michigan would emerge as the first state to ban DDT. And there's more to this story: the connection with the Great Lakes fish population.

Fish Consumption Advisories

In the early 1970s, a lawsuit in Wisconsin sought to ban DDT.[1] Scientists learned that fish oil and fish meal developed from the alewife populations of Lake Michigan and fed to mink on Wisconsin farms resulted in the mink's reproductive failure. Think back to the efforts of the U.S. Bureau of Commercial Fisheries—Dr. Fenton Carbine and other voices—calling for the commercial harvest of alewives for fish meal and fish oil. Dr. Ralph MacMullan was active as a witness in this historic case. He proceeded to participate, choosing to ignore the orders from his governor.

Soon after the case began, Fish Division staff members discovered that coho salmon spawning for the first time had a heavy body burden of DDT.[2] This substance at this concentration resulted in the production of eggs with a higher mortality rate than salmon eggs from other locations. Salmon and other fishes, particularly trout, were found to have dangerously high levels of DDT.

This news came following the excitement that the new fishery had generated. So, it became necessary to advise anglers about minimizing the risk that they might experience when eating their catch.[3] The fish advisories cautioned particularly vulnerable people—young children and women of childbearing age—to limit the frequency of their consumption. Because this contaminant was concentrated in the fish's fatty tissue, government authorities, based on MSU research, advised cleaning, filleting, and trimming carefully to remove as much contaminated fat as possible. They also recommended cooking methods to allow any remaining fat to drip off.

The people of Michigan displayed their stewardship by demanding and achieving the total ban of DDT in 1969—the first state to do so.[4] In due course, in 1972, the federal government banned DDT in all fifty states. The level of DDT in fish and fish eggs has dropped dramatically, and while it still exists at very low levels, it is no longer the problem it once was. Today, a very large number of organizations, including ones that concentrate on conservation and environment, support action and legislation by both the federal government and the various states to protect, enhance, and improve the Great Lakes as a functioning, healthy ecosystem.

Economics

I am neither an economist nor a sociologist, and I am deeply involved personally, so this will probably be judged as biased. For these and other reasons, I will not deal much with precise figures or estimates, but simply remind the reader of the immense economic impact that the development of the Great Lakes sport fishery has had in many dimensions.

Expanded Tourist Season

Much of Michigan's very important summer tourist season is closely associated with activities in Great Lakes shoreline areas. Prior to 1967, the tourist season in these areas generally was considered to begin on Memorial Day and to close following the Labor Day weekend. Suddenly, this season and its profits were extended by two months, and a new industry was being born. That shoreline became known as the Coho Coast or the Gold Coast.[1]

One estimate was that approximately 150,000 anglers were attracted to the fishing in Manistee in September 1967 and spent almost $2 million in the area. "Using the estimate that the tourist dollar turns over three and one-half times before it leaves an area, that is an increase of some $5,850,000," according to William Chasteen (project supervisor for research).[2] Others estimated a $5.2 million economic impact. By October 1968, the Northern Great Lakes Resource Development Committee, represented by Tom Lewis (manager of the Manistee Chamber of Commerce), claimed a $5.8 million impact in that locale.[3]

The Great Lakes, as areas for recreational fishing, have some similarities to tourism in our national parks. All kinds of supporting services—food, lodging, gasoline—need to exist in the communities along the lakeshores. These communities are like such familiar places as the "entrance communities" of Estes Park, Colorado (for Rocky Mountain National Park), and Cody, Wyoming (the town adjacent to Yellowstone National Park). The communities along the Lake Michigan lakeshore—Empire, Frankfort, Manistee, Pentwater—suddenly had many more customers coming through.

Taxes

Several economists in Michigan took a quick look at the sales tax revenues from the areas impacted by the salmon fishery in 1967. That year and the year following saw a clear and dramatic increase in all forms of business activity in the affected communities, and not just a very large boost in sales. Taxable sales in the Manistee area were one million dollars higher than in 1967 and two million higher than in 1966.[4]

The spectacular success of the coho salmon fishery was quite probably involved in creating increases in the value of real estate along the western shore of Michigan's Lower Peninsula. Total assessed valuation in some coastal townships more than doubled between 1966 and 1967. Values jumped again when dead alewives were absent from the beaches during the summer of 1968. By 1970, some of the increases were even larger.[5]

This expanded attractiveness of riparian properties was probably part of such public developments as the Sleeping Bear Dunes National Lakeshore. The final dunes lakeshore boundary is very close to the actual site of the first coho salmon introduction.

Theodore Karamanski, professor of history at Loyola University (Chicago), described it in these terms: "In 1967, the mature fish returned to the river to spawn. Larger than anything most Midwestern fisherman had ever caught, the salmon ignited an acute case of angling fever. The press dubbed the result 'coho madness.'[6] To the manager of Benzie State Park, the rush of hundreds of vehicles and the swarm of anxious anglers did indeed look like madness."[7]

Infrastructure

The exploding salmon fishery quickly revealed something that we had, in a general sense, overlooked. To support the safety and well-being of participants, as well as the ensuing economic development in coastal communities, the location and the adequacy of places to launch and return boats definitely needed urgent attention. The tragic storm of late September 1967 emphatically demonstrated the need and the responsibility to provide safe havens. Similar needs could be anticipated as fishing developed on Lakes Huron, Superior, and Erie.

Within the Department of Conservation, Waterways Division chief Keith Wilson and his staff, whose previous efforts had been concentrated almost exclusively on inland waters, quickly developed a plan for expanding Great Lakes boating access. A key point to this plan was that safe refuge should be available such that no small boat would ever be more than fifteen shoreline miles from safety.[8]

Boating access could be constructed relatively easily and cheaply at many existing harbors. In other areas, they needed to be created. Of course, the key issue was financing. A plan was developed to secure funding from gasoline excise taxes collected by the state. These taxes were generally earmarked for development and maintenance of highways. However, it was logical that gasoline used by boats should be a source to support facilities for boating. A formula was developed based on the percentage of gasoline used by boats within the total amount of gasoline sold. With some delay, the legislature recognized the need and the logic, and these sources of tax revenue ultimately became available to the division. Michigan can be proud of the outstanding facilities now available for small boats. The facilities where boats can be docked for the season and other areas where small boats can be launched have become one of the best systems in the nation. Other states have followed Michigan's example.

In time, many varied arrangements for building, maintenance, and management were developed between the Waterways Division, elements of local governments, and other community interests.

Lures and tackle

MICHIGAN DEPARTMENT OF CONSERVATION/
NATURAL RESOURCES

Gear

Rods, reels, downriggers, cannonballs, fish finders, radios, landing nets, life jackets, and millions of fishing lures—gear of many types were needed. By October 1969, in a presentation to the Michigan Natural Resources Commission, Don Elwood of the American Fishing Tackle Manufacturers Association said that the introduction of Pacific salmon to the Great Lakes was the "greatest shot in the arm that sport fishing in the U.S. has received." It was valued at more than $150 million in Michigan in 1968. The retail value of fishing tackle shipped by Michigan factories jumped from $35 million in 1963 to more than $60 million in 1968. Helin Tackle Company (Detroit) had a net gain in sales from $82,000 in 1966 to $355,000 in 1968. A representative of Shakespeare in Kalamazoo said, "No question but what the salmon and trout program in the Great Lakes is by far the largest single fishing development in the history of our industry."[9]

Boats and Motors

For many years, Michigan registered more boats than any other state except California, and a significant percentage of them were used for fishing.¹⁰ To put this in perspective, California has 38 million people, while we have fewer than 10 million. Our fishermen all needed bigger boats to participate in this fishery; nothing less than sixteen feet would do, and most were twenty to twenty-two feet, with some of the larger boats going thirty to thirty-six feet long, with larger outboard motors.

In all my forty some years prior to the development of this fishery, I never saw more than one or two outboard motors as large as 25 hp. Suddenly, fishing boats required big ones to be safe and to be able to travel the long distances that fishermen would go on the Great Lakes. A large outboard was mandatory; for safety purposes, each boat should have two, in case one failed. Twenty-five horsepower would be minimal, and 50, 75, 125, and even 250 hp outboard motors were used. I can't put numbers on the size of this market, but it certainly represented a monumental increase in manufacturing and sales opportunities.

Boat names are a form of self-expression, and I knew someone who named his boat SHENEEDA; he would say, "She need a new battery, she need a new prop or a new radio, etc., etc." This expression reminds me of the definition of a boat—a hole in the water into which you pour money.

Jobs

A very large number of fishing-related jobs developed—from charter boat captains to those who help manufacture and sell equipment, to those in various service functions. For example, charter fishing is an important contributor to tourism in Michigan's coastal communities. A 2009 study by Michigan Sea Grant and MSU's Center for Economic Analysis found that, over the preceding twenty years, charter fishing had generated an average of 465,417 employment hours per year and brought an average of $19.8 million per year into Michigan's coastal communities.¹¹

Events

Another example is fishing tournaments, which have become quite popular. The Village of Honor even developed a "National Coho Festival" (the first of its kind).[12] It now usually happens the fourth weekend in August, celebrating the "salmon run" up the Benzie County rivers. Honor is considered the "birthplace" of salmon in the state of Michigan, because of its proximity to the Platte River, where the first salmon were planted.

Sixteen events for trout and salmon occurred on the Lake Michigan shoreline in 2009. The three-day Grand Haven Salmon Festival has included live music, wine and salmon tasting, and children's activities such as a salmon-fishing simulator. Recreational fishing provided a focal point for the festival, but most economic impacts were generated by 1,061 tourists who traveled to participate in festival events but did not participate in the Big King Contest.[13]

The economic value of the Great Lakes recreational fishery, according to a frequently used estimate, says "Michigan's Great Lakes sport fishery generates an economic value of $4–8 billion per year!"[14] This fishery has been ours to enjoy for more than fifty years.

CONCLUSION

ore than fifty years have passed since we successfully introduced Pacific salmon into Michigan waters of the Great Lakes. A lot has happened in the world during those decades. Humans have landed on Earth's moon and continue to orbit our "big blue marble" in an international space station. Even from there they can recognize and observe our Great Lakes. Closer to home, the lakes have been the focus of international water quality agreements and federal and interstate cooperative fisheries management, as well as state and provincial initiatives.

Communications and consultations once happened by telephone, "snail" mail, the ponderous process of scientific publication, and face-to-face conferences. We now can transmit messages instantaneously, and I'm able to continue to "read" and "write" through electronic devices even though my eyesight has deteriorated significantly. We have much faster, easier access to the results of scientific research.

I haven't been directly responsible for fisheries management since 1966, but I served for several years as director of Michigan's Department of Natural Resources, overseeing the Fisheries Division and others. I have also maintained a consistent and keen interest in our state's fishery, especially in the Great Lakes.

It's been an amazing experience.
MICHIGAN DEPARTMENT OF CONSERVATION/NATURAL RESOURCES

Today?

As I look back at all the hurdles that we had to successfully surmount to release the salmon into the Great Lakes, I am amazed that we succeeded, and, in response to occasional questions, I wonder what would happen if that opportunity arose today. I doubt that it would happen. The state now has a ban on the introduction of new species, which is intended to prevent the transmission of viral hemorrhagic septicemia (VHS), bacterial kidney disease (BKD), and other fish diseases. An Environmental Impact Statement would be required, and I doubt that, with such a statement, we would have received approval.

That said, in looking back over a lifetime of experiences dealing, in one way or another, with the biology of our freshwater environments, I can see that constant change and response are integral, imperative aspects of policy, research, and management. That's what makes it so exciting. Over these years, I've observed several important developments—some that give me great concern, and some that give me hope for the future of this magnificent fishery.

Fifty years ago, Pacific salmon took to the Great Lakes easily and helped establish important redundancy in our top predators. Natural reproduction of chinook and coho has augmented our "put and take" hatchery-based system of management. We took advantage of their natural appetite for the invasive alewife, which lived throughout the pelagic zone of the water column. In recent years, this food source has declined because they have had to compete for planktonic nutrition with quagga mussels.

Until recently, the round gobi has been considered an unwanted and unfortunate invasive species. However, it is now feeding on the superabundant quagga mussels and is an available food source for all the predaceous trout and salmon that can adapt to feeding in the benthic zone. It's exciting and a challenge for fisheries professionals to adjust their management policies and strategies. These salmonids have also endured diseases, but fishery managers have surmounted those difficulties, too, and anglers have continued to enjoy Great Lakes sport fishing.

The U.S. and Canadian governments have managed to keep the voracious sea lamprey under control with a variety of techniques. However, that is a very expensive undertaking that controls but will never eliminate them, and I wonder if we will continue to invest our tax dollars in this endeavor. Alewife populations have fluctuated wildly, as might be expected, and the salmon have become opportunistic in foraging for their prey.

Besides that, we have experienced some extremely troublesome and not yet resolved invasions of other species. Zebra and quagga mussels, especially the quaggas, have destroyed habitat and made major incursions into the Great Lakes food web. Spiny water fleas have proved to be nutritionally inadequate and disruptive to diets of younger fishes. Round gobies tend to prey on the young of other species. Asian carps of several species are poised on the threshold of the Great Lakes Basin, if they haven't already succeeded in reaching this desirable (for them) destination. It's important to learn all that we can about these species' native habitats and do everything that we can to minimize their negative impact on our incomparable system.

It is even more important that we be proactive and protective in considering developments that can create additional havoc in this already challenged environment. For example, we already know that numerous invasive species have made their way into the Great Lakes ecosystem via oceangoing vessels through the St. Lawrence Seaway. Why enlarge that system or continue to permit such vessels to discharge their ballast here? And why are urban water-treatment facilities allowed to fail their critical sanitary function and be overwhelmed by storm water, flushing millions of gallons of untreated sewage into "fresh" drinking water? Or why are we so consumed by our "need" for oil and gas that we ignore or accept the risk of aging pipelines crisscrossing our beautiful peninsulas and traversing our Great Lakes, leaving us so exposed to serious, potentially deadly pollution incidents? Why do we think it might be a good idea to allow some areas of the Great Lakes to be farmed for fish (in pens that will concentrate their waste) that will compete with other species?

More Support Needed

I've also observed a significant decline in the adequacy of support for Michigan's fisheries management staff. Consider this: Approximately 50 percent of all the surface acres of fresh water, lakes, and streams in this nation's fifty states—including Alaska—lies within the boundaries of Michigan. Please think about it! Fifty percent of all waters of this nation are within Michigan, making us as citizens and, perhaps more importantly, future citizens responsible for their perpetual care. We must provide the policies, rules, regulations that will keep these waters clean and productive. Of course, the Fisheries Division has the responsibility to manage the fish populations in a manner that provides for their care, and to enforce the rules that regulate how we participate in and enjoy this fishery resource.

However, our Fisheries Division is currently woefully underfunded—understaffed and underpaid—when contrasted with the size of its responsibilities. A conservative average of personnel in the fisheries divisions of other states is 200 people. In Michigan, our Fisheries Division, with the responsibility of 50 percent of all the waters in this country, has a total staff of about 150 people. If you take that average of 200 for the remaining forty-nine states, you come up with a figure of nearly 10,000 people managing one half of the total resource, while 150 people manage Michigan's half of the resource!

Great Lakes Fisheries Leadership

The people of Michigan must accept our state's responsibility for Great Lakes leadership, especially in fisheries. We must have state leaders who are committed to good natural-resource and environmental policies in the management of our precious public resources. We need elected officials who consider the long-term implications of decisions involving the Great Lakes ecosystem and who won't sell us short for temporary economic gains. We need citizens who take their roles seriously, electing and holding their representatives accountable to preserving these "pleasant peninsulas" for generations to come. We need taxpayers who are willing to invest in the long-term sustainability of the Great Lakes fishery.

Professional Leadership

We need highly professional resource managers willing to use all their knowledge, expertise, and skills on behalf of these Great Lakes. Those managers need the best tools, equipment, and facilities to do their job. Fisheries management should be about 50 percent dealing with fish and 50 percent dealing with people, and we could be doing so much more if we had the financial support and, of course, the staff. If more people were available, I could suggest a dozen places of interactions with the fishing public and lakeside property owners on how to attempt to enhance various aspects of fishing opportunity.

We need some new facilities to improve Great Lakes management. Future fishery managers will be scrutinized much more significantly than we were when we brought coho and chinook salmon to the lakes. Therefore, we need a diagnostic and quarantine center that will allow resource managers to examine and evaluate additional races and species of fishes that might be introduced into the Great Lakes ecosystem. I think that Atlantic salmon could certainly play an important role in the sport fishery here, and we need to expand our efforts in that direction.

Citizen Leadership

Most importantly, we need the public to understand that roughly 40 percent of the surface area of Michigan is in the Great Lakes. We must teach that this produces income, and having taught the public, they must convince the legislature to adequately fund the instruments of management, enhancement, and stewardship.

I know that those who have followed me in managing Michigan's fisheries have worked hard to sustain viable sport and commercial fishing opportunities and have continued to learn from our initial experience and to engage fishing folks throughout the state. Without them, we wouldn't continue to enjoy the fine fishery that we currently have. I hope that those of you who have read this far will support them in their efforts and become advocates, if you aren't already, so that generations to come will also be able to fish for these spectacular salmon.

Youth Concerns

I'm also concerned about our young people being so absorbed with their technological devices, and their lack of involvement with the outdoors. Who will be fishing in the future? Who will value and enjoy our state's "water wonderland" and "Great Lakes splendor"? Who will bring together the variety of small, narrowly focused conservation and environmental organizations to address the broader issues that affect everyone's well-being? We need younger people to be educated about, get seriously involved with, and become leaders in advocating for the resources that they may tend to take for granted.

Whether through school-based fishing programs, summer camps, or youth memberships in fishing organizations, we must engage young people with the outdoors and the environment. We must encourage them to pursue careers in fisheries management and to see that, whatever their occupation or position, they have an important impact on the environment for better or worse, and that it will affect their future, too.

The Future

Who will be fishing for salmon in the Great Lakes fifty years from now? Time will tell, of course, but I do hope that succeeding generations will value what we started some fifty years ago—the largest and most valuable freshwater sport fishery in the world—and will make the commitment necessary to keep it great.

Is Great Lakes fishing in their future?

NOTES

INTRODUCTION

1. American Fisheries Society, Howard A. Tanner, Fisheries Management Section, Hall of Excellence, 2008, http://fms.fisheries.org/awards/hall-of-excellence/virtual-hall-of-excellence/.

2. Al Spiers, "Miracle of the Fishes," *Saturday Evening Post*, Fall 1972.

3. Ronald Schiller, "Michigan: The State That Almost Wasn't, Armchair Travelogue," *Reader's Digest*, February 1975.

4. Southwick Associates, *Sportfishing in America: An Economic Force for Conservation*, produced for the American Sportfishing Association (ASA) under a U.S. Fish and Wildlife Service Sport Fish Restoration grant (F12AP00137, VA M-26-R), awarded by the Association of Fish and Wildlife Agencies, 2012.

GROWING UP WITH THE FISHES

1. "Jan Metzlaar (1891–1929): Fisheries Research Biologist," in Michigan Department of Natural Resources, *Fisheries Management Report No. 6, Michigan Fisheries Centennial Report—1873–1973, Fisheries Division—April 1974*.

2. "Howard Tanner, Sheriff, Antrim County," *Traverse City Record-Eagle*, December 31, 1960.

THE "GREATEST GENERATION" GOES TO WAR

1. "Detailed Tables: 51 Population of Counties, Incorporated Places and Minor Civil Divisions," in *Census of Population and Housing, 1920* (Washington, DC: United States Census Bureau, 1920), 232–33, https://www.census.gov/prod/www/decennial.html.
2. Tom Brokaw, *The Greatest Generation* (New York: Random House, 2004).
3. Louis E. Keefer, *Scholars in Foxholes: The Story of the Army Specialized Training Program in World War II* (Jefferson, NC: McFarland & Co., 1988).
4. Marine Lieut. Ellis M. Trefethen, "This Is Tarawa Today," *New York Times Magazine*, March 19, 1944, quoted in "Golden Gate in '48," *Time*, March 27, 1944, "World Battlefronts: On to Westward."
5. U.S. Army Center for Military History, Luzon 1944–1945, http://www.history.army.mil/brochures/luzon/72-28.htm.

BUILDING THE EDUCATIONAL FRAMEWORK

1. Paul S. Welch, *Limnological Methods* (Philadelphia: Blakiston Co., 1948).
2. I. B. Bird, Obituaries, *Wildlife Society Bulletin (1973–2006)* 1, no. 3 (Autumn 1973): 157–59, http://www.jstor.org/stable/i292147.
3. W. R. Crowe, "Activities of Fisheries Biologist, 1947, District No. III," Fisheries Research Report 1157, http://quod.lib.umich.edu/f/fishery.
4. M. L. Vitosh, "NPK Fertilizers," Michigan State University Extension Bulletin E-896, reprint July 1996, http://fieldcrop.msu.edu/uploads/documents/e0896.pdf.
5. Howard A. Tanner, "The Biological Effects of Fertilizer on a Natural Lake," PhD diss. (Michigan State College of Agriculture and Applied Science, Department of Zoology, 1950).
6. Howard A. Tanner, "Some Consequences of Adding Fertilizer to Five Michigan Trout Lakes," *Transactions of the American Fisheries Society* 89, no. 2 (1960): 198–205.
7. Frank F. Hooper, R. C. Ball, and H. A. Tanner, "An Experiment in the Artificial Circulation of a Small Michigan Lake, *Transactions of the American Fisheries Society* 82, no. 1 (1953): 222–41.

PROFESSIONAL PRACTITIONER

1. U.S. Geological Survey, "History of the Colorado Unit," Cooperative Research Units, 2015, http://www.coopunits.org/Colorado/History/.
2. *Columbia Electronic Encyclopedia*, 6th ed. (New York: Columbia University Press), http://

www.infoplease.com/encyclopedia/us/colorado-state-united-states-geography.html.

3. Ibid.

4. "Detailed Tables: Population of Counties, Incorporated Places and Minor Civil Divisions," in *Census of Population and Housing, 1950* (Washington, DC: United States Census Bureau), https://www.census.gov/prod/www/decennial.html.

5. Michigan Department of Environmental Quality, "Water: Great Lakes," http://www. michigan.gov/deq/0,4561,7-135-3313--,00.html.

6. Water Information Program, "Colorado Water Rights," http://www.waterinfo.org/rights. html.

7. U.S. Department of the Interior, Bureau of Reclamation, Upper Colorado Region, "Colorado River Storage Project," http://www.usbr.gov/uc/rm/crsp/.

8. Colorado Parks and Wildlife, "Brook Trout Removal Helps Stabilize Rio Grande Cutthroat," http://cpw.state.co.us/aboutus/Pages/News-Release-Details. aspx?NewsID=5484.

9. National Park Service, Mississippi, National River and Recreation Area, "History of Common Carp in America," http://www.nps.gov/miss/learn/nature/carphist.htm.

10. Colorado Fisherman.com, "Fish Species in Colorado Waters," http://www. coloradofisherman.com/fish_species_of_colorado.php.

11. W. D. Klein, *Kokanee in Parvin Lake, Colorado, 1972–1977* (Denver: Colorado Division of Wildlife, 1979).

12. Colorado Parks and Wildlife, "Fish Species Identification," http://cpw.state.co.us/learn/ pages/fishid.aspx.

13. W. D. Klein and L. M. Finnell, "Comparative Study of Coho Salmon Introductions in Parvin Lake and Granby Reservoir," *Progressive Fish-Culturist* 31, no. 2 (1969), http://www. tandfonline.com/.

14. Delbert A. West, "Freshwater Silver Salmon, *Oncorhynchus kisutch* (Walbaum)," *Calif. Dept. Fish & Game* 51 (1965): 210–12, in Wayne H. Tody and Howard A. Tanner, Fish Division, *Coho Salmon for the Great Lakes*, Fish Management Report No. 1, February 1966.

15. Arthur Whitney, personal communication, 1965, Montana Department of Fish and Game, Helena, in Wayne H. Tody and Howard A. Tanner, Fish Division, *Coho Salmon for the Great Lakes*, Fish Management Report No. 1, February 1966. P. Fuller, J. Larson, and A. Fusaro, "*Oncorhynchus kisutch*," USGS Nonindigenous Aquatic Species Database, Gainesville, FL, https://nas.er.usgs.gov/queries/FactSheet.aspx?speciesID=908 (rev. 6/26/2014).

16. Klein and Finnell, "Comparative Study."

THE INLAND SEAS

1. A. M Beeton, C. E. Sellinger, and D. F. Reid, "An Introduction to the Laurentian Great Lakes Ecosystem," in *Great Lakes Fisheries Policy and Management: A Binational Perspective*, ed. W. W. Taylor and C. P. Ferreri (East Lansing: Michigan State University Press, 1999), 3–54.

2. Kent Fuller, Harvey Shear, and Jennifer Wittig, eds., *The Great Lakes: An Environmental Atlas and Resource Book*, 3rd ed. (Toronto/Chicago: Government of Canada and U.S. Environmental Protection Agency, 1995).

3. Ibid.

4. Coordinating Committee on Great Lakes Basic Hydraulic and Hydrologic Data, *Coordinated Great Lakes Physical Data, May 1977* (Detroit: U.S. Army Corps of Engineers, 1977).

5. Fuller, Shear, and Wittig, *The Great Lakes*.

6. Ibid.

7. Southwick Associates, https://www.southwickassociates.com.

8. U.S. Department of the Interior, U.S. Fish and Wildlife Service and U.S. Department of Commerce, *2011 National Survey of Fishing, Hunting and Wildlife-Associated Recreation*, https://www.census.gov/prod/2012pubs/fhw11-nat.pdf.

HUMAN HISTORY AND THE GREAT LAKES FISHERY

1. Matthew McCarthy, *Native American Population Decline during the Nineteenth Century*, https://nativestudy.wordpress.com; U.S. Army Corps of Engineers, "Treaty Rights and Subsistence Fishing in the U.S. Waters of the Great Lakes, Upper Mississippi River, and Ohio River Basins," *Great Lakes and Mississippi River Interbasin Study*, 2012, http://glmris. anl.gov/documents/docs/Subsistence_Fishing_Report.pdf; Department of Geography, Environment, and Spatial Sciences, Michigan State University, "Native Americans in the Great Lakes Region," http://geo.msu.edu/extra/geogmich/paleo-indian.html; Margaret Beattie Bogue, *Fishing the Great Lakes: An Environmental History, 1783–1933* (Madison: University of Wisconsin Press, 2001); "Fishing in the Western Great Lakes Region," Native American Netroots, http://nativeamericannetroots.net/diary/1341.

2. Ontario Federation of Anglers and Hunters, "Atlantic Salmon Early History," Bring Back the Salmon—Lake Ontario, 2016, http://www.bringbackthesalmon.ca/biology-history/ early-history/.

3. Ontario Federation of Anglers and Hunters, "Extirpation and Early Recovery Efforts," Bring Back the Salmon—Lake Ontario, 2016, http://www.bringbackthesalmon.ca/ biology-history/extirpation-early-recovery-efforts/.

4. Tom Kuchenberg, *Reflections in a Tarnished Mirror: The Use and Abuse of the Great Lakes* (Sturgeon Bay, WI: Golden Glow Publishing, 1978).

5. Dave Dempsey, *Ruin and Recovery: Michigan's Rise as a Conservation Leader* (Ann Arbor: University of Michigan Press, 2001).

6. John W. Parsons, *History of Salmon in the Great Lakes, 1850–1970*, Technical Paper 68 (Washington, DC: U.S. Department of the Interior, Fish and Wildlife Service, Bureau of Sport Fisheries and Wildlife, April 1973).

7. "*Oncorhynchus mykiss,*" USGS Nonindigenous Aquatic Species Database, Gainesville, FL, and NOAA Great Lakes Aquatic Nonindigenous Species Information System, Ann Arbor, MI, https://nas.er.usgs.gov/queries/greatlakes/FactSheet. aspx?SpeciesID=910&Potential=N&Type=0&HUCNumber=DGreatLakes (rev. 11/4/2013); "*Salmo trutta,*" USGS Nonindigenous Aquatic Species Database, Gainesville, FL, and NOAA Great Lakes Aquatic Nonindigenous Species Information System, Ann Arbor, MI, https://nas.er.usgs.gov/queries/greatlakes/FactSheet. aspx?SpeciesID=931&Potential=N&Type=0&HUCNumber=DGreatLakes (rev. 1/6/2015).

8. "*Osmerus mordax,*" USGS Nonindigenous Aquatic Species Database, Gainesville, FL, https://nas.er.usgs.gov/queries/FactSheet.aspx?speciesID=796 (rev. 9/29/2015).

9. Thomas G. Coon, "Ichthyofauna of the Great Lakes Basin," in *Great Lakes Fisheries Policy and Management: A Binational Perspective*, ed. W. W. Taylor and C. P. Ferreri (East Lansing: Michigan State University Press, 1999), 55–71.

10. U.S. Bureau of Commercial Fisheries, *A Century of Fish Conservation (1871–1971)*, http://nctc.fws.gov/History/Articles/FisheriesHistory.html.

FISHERIES MANAGEMENT

1. U.S. Department of Commerce, National Oceanic and Atmospheric Administration, *Fisheries Glossary*, 2006 rev. ed., https://www.st.nmfs.noaa.gov/st4/documents/ FishGlossary.pdf.

2. Federation of European Aquaculture Producers (FEAP), "Introduction," http://www.feap. info/default.asp?SHORTCUT=589.

3. William F. Royce, "The Historical Development of Fisheries Science and Management," taken from a lecture given at the Fisheries Centennial Celebration (1985), http://www. nefsc.noaa.gov/history/stories/fsh_sci_history1.html.

4. U.S. Geological Survey, Great Lakes Science Center, http://www.glsc.usgs.gov/library/.

MEANWHILE IN MICHIGAN

1. State of Michigan, Legislature, *Michigan Constitution*, http://www.legislature.mi.gov/ documents/Publications/Constitution.pdf.

2. McLaughlin et al., *Governor's Special Conservation Study (Blue Ribbon) Committee*, presented to Governor Romney, Lansing, MI, 1963; Wildlife Management Institute, *Report to the Governor's Special Conservation Study Committee on the Michigan Department of Conservation*, 1963.

3. G. F. Whelan, "A Historical Perspective on the Philosophy behind the Use of Propagated Fish in Fisheries Management: Michigan's 130 Year Experience," *American Fisheries Society Symposium* 44 (2004): 307–15, https://www.michigan.gov/documents/dnr/ Historical-Perspective-on-use-of-Hatchery-Fish_226808_7.pdf.

4. Ibid.

5. Howard A. Schuck, "Survival of Hatchery Trout in Streams and Possible Methods of Improving the Quality of Hatchery Trout," *Progressive Fish-Culturist* 10, no. 1 (1948), https://doi.org/10.1577/1548-8640(1948)10[3:SOHTIS]2.0.CO;2.

6. W. C. Latta, "Early History of Fisheries Management in Michigan," *Fisheries* 31, no. 5 (2006): 230–34.

7. U.S. Geological Survey, Great Lakes Science Center, http://www.glsc.usgs.gov/library/.

8. Michigan Department of Conservation, Fish Section, *A Management Program for Michigan's Fisheries* (McFadden Report), 1964.

9. Michigan Department of Conservation, "Proceedings of Conservation Commission, June 12, 1964, Lansing, Michigan," Digital DNR, http://cdm15867.contentdm.oclc.org/cdm/ singleitem/collection/p15867coll4/id/1312/rec/1.

COMMERCIAL FISHING

1. Vernon C. Applegate and H. D. Van Meter, *A Brief History of Commercial Fishing in Lake Erie*, Fishery Leaflet 630 (Washington, DC: United States Department of the Interior, April 1970).

2. Tom Goniea, "The Story of State-Licensed Commercial Fishing History on the Great Lakes," http://www.michigan.gov/dnr/0,4570,7-153-10364_52259-316019--,00.html.

3. Michigan Department of Natural Resources, Fisheries Division, Fisheries Management Report No. 6, April 1974; *Michigan Fisheries Centennial Report: 1873–1973*, http://www. michigandnr.com/publications/pdfs/ifr/ifrlibra/special/reports/sr06.pdf.

4. D. A. Brege and N. R. Kevern, *Michigan Commercial Fishing Regulations: A Summary of Public Acts and Conservation Commission Orders, 1865 through 1975*, Michigan Sea Grant College Program, MICHU-SG-78-605 (East Lansing: Michigan State University, 1978).

5. International Pacific Halibut Commission, *Convention for the Preservation of the Halibut Fishery of the Northern Pacific Ocean and Bering Sea, 1953* (TIAS 2900), http://www.nmfs.noaa.gov/ia/agreements/regional_agreements/pacific/iphc.pdf.

THE GREAT LAKES FISHERY COMMISSION

1. William R. Willoughby, *The Joint Organizations of Canada and the United States* (Toronto: University of Toronto Press, 1979), 77–78; Margaret R. Dochoda and M. L. Jones, "Managing Great Lakes Fisheries under Multiple and Diverse Authorities," in *Sustaining North American Salmon: Perspectives across Regions and Disciplines*, ed. Kristine D. Lynch, M. L. Jones, and W. W. Taylor (Bethesda, MD: American Fisheries Society, 2002), 221–42. Michael J. Donahue, *A Detailed Review of Selected Institutions for Great Lakes Management*, "Appendix: Institutional Arrangements for Great Lakes Management: Past Practices and Future Alternatives," Michigan Sea Grant College Program, November 1987, MICHU-SG-87-200T (Ann Arbor: University of Michigan 1987).
2. Dochoda and Jones, "Managing Great Lakes Fisheries."

TRIBAL FISHING RIGHTS

1. "Big Abe" and "The Indian Rights Issue: Fishing," *Detroit Free Press*, January 6, 1980.
2. Native American Rights Fund, *Bay Mills Indian Community, et al. v. The State of Michigan* (*hunting and fishing*), http://www.narf.org/our-work/case-map/.
3. Great Lakes Indian Fish and Wildlife Commission, "Treaty Rights," http://www.glifwc.org/TreatyRights/.
4. Michigan's 1836 Treaty Fishery Guide—Abe LeBlanc 1971, http://www.glifwc.org/publications/.

LAKE TROUT

1. P. Fuller and M. Neilson, "*Salvelinus namaycush*," USGS Nonindigenous Aquatic Species Database, Gainesville, FL, https://nas.er.usgs.gov/queries/FactSheet.aspx?speciesID=942 (rev. 2/2/2016).
2. University of Michigan Museum of Zoology, "Animal Diversity Web," http://animaldiversity.ummz.umich.edu/accounts/Salvelinus_namaycush/#geographic_range.
3. U.S. Fish and Wildlife Service, "Jordan River National Fish Hatchery," 2013, http://www.fws.gov/midwest/jordanriver/about.html.
4. State of Michigan, Department of Natural Resources, *Marquette State Fish Hatchery, 1922–1972*, http://www.michigan.gov/documents/dnr/History1922-1972_288374_7.pdf.

SEA LAMPREY

1. Great Lakes Fishery Commission, "Sea Lamprey: A Great Lakes Invader," http://www. glfc.org/sea-lamprey.php; P. Fuller, L. Nico, E. Maynard, J. Larson, A. Fusaro, and A. K. Bogdanoff, "*Petromyzon marinus,*" USGS Nonindigenous Aquatic Species Database, Gainesville, FL, and NOAA Great Lakes Aquatic Nonindigenous Species Information System, Ann Arbor, MI, https://nas.er.usgs.gov/queries/factsheet.aspx?SpeciesID=836 (rev. 9/29/2016).

2. V. C. Applegate, "Natural History of the Sea Lamprey (*Petromyzon marinus*) in Michigan," U.S. Department of the Interior, Fish and Wildlife Service, Fisheries Research Report 1254.

3. Fuller, Nico et al., "*Petromyzon marinus.*"

4. J. Ellen Marsden and R. W. Langdon, "The History and Future of Lake Champlain's Fishes and Fisheries," *Journal of Great Lakes Research* 38 (2012): 19–34.

5. John W. Parsons, *History of Salmon in the Great Lakes, 1850–1970*, Technical Paper 68 (Washington, DC: U.S. Department of the Interior, Fish and Wildlife Service, Bureau of Sport Fisheries and Wildlife, April 1973).

6. Great Lakes Fishery Commission, "Sea Lamprey: A Great Lakes Invader"; M. Siefkes and J. Wingfield, "Understanding Sea Lamprey: Mapping the Genome and Identifying Pheromones," Great Lakes Fishery Commission, 1950, http://www.glfc.org/eforum/article1.html.

7. Michigan Department of Conservation, *Official News Bulletin*, December 18, 1964.

ALEWIFE

1. P. Fuller, E. Maynard, D. Raikow, J. Larson, A. Fusaro, and M. Neilson, "*Alosa pseudoharengus,*" USGS Nonindigenous Aquatic Species Database, Gainesville, FL, https://nas.er.usgs.gov/queries/FactSheet.aspx?speciesID=490 (rev. 9/25/2015).

2. Ibid.

3. Ibid.

4. Lynn Ceci, "Fish Fertilizer: A Native North American Practice?" *Science*, n.s., 188, no. 4183 (April 4, 1975): 26–30.

5. Fuller, Maynard et al., "*Alosa pseudoharengus.*"

6. Michigan Department of Natural Resources, "History of State-Licensed Great Lakes Commercial Fishing," http://www.michigan.gov/dnr/0,4570,7-350-79136_79236_80538_80541-424724--,00.html.

SALMON

1. P. Fuller, J. Larson, and A. Fusaro, "*Oncorhynchus kisutch*," USGS Nonindigenous Aquatic Species Database, Gainesville, FL, https://nas.er.usgs.gov/queries/FactSheet.aspx?speciesID=908 (rev. 6/26/2014). Also John W. Parsons, *History of Salmon in the Great Lakes, 1850–1970*, Technical Paper 68 (Washington, DC: U.S. Department of the Interior, Fish and Wildlife Service, Bureau of Sport Fisheries and Wildlife, April 1973).

2. P. Fuller, J. Liebig, J. Larson, and A. Fusaro, "*Oncorhynchus gorbuscha*," USGS Nonindigenous Aquatic Species Database, Gainesville, FL, https://nas.er.usgs.gov/queries/FactSheet.aspx?SpeciesID=906 (rev. 6/26/2014).

3. William R. Heard, "Life History of Pink Salmon," in *Pacific Salmon Life Histories*, ed. C. Groot and L. Margolis (Vancouver: University of British Columbia Press, in cooperation with the Government of Canada, Department of Fisheries and Oceans, 1991).

4. P. Fuller, G. Jacobs, J. Larson, T. H. Makled, and A. Fusaro, "*Oncorhynchus nerka*," USGS Nonindigenous Aquatic Species Database, Gainesville, FL, https://nas.er.usgs.gov/queries/factsheet.aspx?SpeciesID=915 (rev. 7/8/2014).

5. Lou Klewer, "Outdoors with Lou," *Toledo Blade*, November 1, 1964.

6. American Fisheries Society, *Common and Scientific Names of Fishes from the United States, Canada, and Mexico* (AFS Special Publication 34; 2013).

7. Michigan Department of Natural Resources, "Chinook salmon *Oncorhynchus tshawytscha*," http://www.michigan.gov/dnr/0,4570,7-153-10364_18958-45663--,00.html; P. Fuller, G. Jacobs, M. Cannister, J. Larson, and A. Fusaro, "*Oncorhynchus tshawytscha*," USGS Nonindigenous Aquatic Species Database, Gainesville, FL, https://nas.er.usgs.gov/queries/FactSheet.aspx?SpeciesID=920 (rev. 6/26/2014).

8. Michigan Department of Conservation, *Official News Bulletin*, December 18, 1964.

A NEW DAY DAWNS

1. W. F. Hublou, J. Wallis, T. B. McKee, D. K. Law, R. O. Sinnhuber, and T. C. Yu, "Development of the Oregon Pellet Diet," *Fish Commission Oregon Res. Briefs* 7, no. 1 (1959): 28; Oregon State University Coastal Oregon Marine Experiment Station, "Innovations and Improvements—Behind the Scenes," 2015, http://marineresearch.oregonstate.edu/innovations-and-improvements-behind-scenes.

SUPPORTING AND OPPOSING FORCES

1. Michigan Department of Conservation, "Proceedings of Conservation Commission, November 13, 1964, Lansing, Michigan," Archives of Michigan, Lansing.

2. J. E. Lasater to Howard A. Tanner, December 15, 1964.

3. Robert W. Schoning to Howard A. Tanner, December 16, 1964.

4. Michigan Department of Conservation, *Official News Bulletin*, December 18, 1964.

5. L. Nico, E. Maynard, P. J. Schofield, M. Cannister, J. Larson, A. Fusaro, and M. Neilson, "*Cyprinus carpio*," USGS Nonindigenous Aquatic Species Database, Gainesville, FL, https://nas.er.usgs.gov/queries/factsheet.aspx?speciesID=4 (rev. 7/15/2015).

6. P. Fuller, L. Nico, E. Maynard, J. Larson, A. Fusaro, and A. K. Bogdanoff, "*Petromyzon marinus*," USGS Nonindigenous Aquatic Species Database, Gainesville, FL, and NOAA Great Lakes Aquatic Nonindigenous Species Information System, Ann Arbor, MI, https://nas.er.usgs.gov/queries/factsheet.aspx?SpeciesID=836 (rev. 9/29/2016).

7. P. Fuller, E. Maynard, J. Larson, A. Fusaro, T. H. Makled, and M. Neilson, "*Osmerus mordax*," USGS Nonindigenous Aquatic Species Database, Gainesville, FL, https://nas.er.usgs.gov/queries/factsheet.aspx?SpeciesID=796 (rev. 9/29/2015).

8. P. Fuller, E. Maynard, D. Raikow, J. Larson, A. Fusaro, and M. Neilson, "*Alosa pseudoharengus*," USGS Nonindigenous Aquatic Species Database, Gainesville, FL, https://nas.er.usgs.gov/queries/FactSheet.aspx?speciesID=490 (rev. 9/25/2015).

9. P. Fuller, J. Larson, and A. Fusaro, "*Oncorhynchus kisutch*," USGS Nonindigenous Aquatic Species Database, Gainesville, FL, https://nas.er.usgs.gov/queries/FactSheet.aspx?SpeciesID=908 (rev. 6/26/2014). Also John W. Parsons, *History of Salmon in the Great Lakes, 1850–1970*, Technical Paper 68 (Washington, DC: U.S. Department of the Interior, Fish and Wildlife Service, Bureau of Sport Fisheries and Wildlife, April 1973).

10. Scott Willoughby, "State of Native Trout in Colorado Is Grim, According to Report," *Denver Post*, June 23, 2015, http://www.denverpost.com/2015/06/23/state-of-native-trout-in-colorado-is-grim-according-to-report/ (rev. 4/24/16).

11. Association for the Sciences of Limnology and Oceanography, 2013, http://www.aslo.net/photopost/showphoto.php/photo/1104/title/rainbow-trout-from-lake-titicaca-2c-peru-and-bolivia/cat/517.

12. Government of the United States of America and Government of Canada, *Convention on Great Lakes Fisheries between the United States and Canada, September 10, 1955*, http://www.glfc.org/pubs/conv.htm.

13. Vernon C. Applegate and H. D. Van Meter, "A Brief History of Commercial Fishing in Lake Erie," United States Department of the Interior, Fishery Leaflet 630, April 1970, https://pubs.usgs.gov/unnumbered/81373/report.pdf.

14. Great Lakes Fishery Commission, *A Prospectus for Investigations of the Great Lakes Fishery*, August 15, 1964.

15. Trout Unlimited, "Who We Are," http://www.tu.org/about-tu.

16. Trout Unlimited, "History," http://www.tu.org/about-tu/history.

17. Montana Trout Unlimited, "A History of TU in Montana," http://montanatu.org/wp-content/uploads/2012/11/MTU-History-to-2009.pdf.
18. Michigan Department of Natural Resources, Forestry, "Grayling: Mason Tract," http://www.michigan.gov/dnr/0,4570,7-153-30301_30505-66206--,00.html.
19. Mark A. Tonello, "Little Manistee River," in *Michigan Department of Natural Resources, Status of the Fishery Resource Report, 2005–08*, https://www.michigan.gov/documents/2005-8_Little-Manistee_River_144067_7.pdf.

EGGS FROM OREGON

1. Leonard N. Allison to Howard A. Tanner, February 5, 1965.
2. Michigan Department of Conservation, *Official News Bulletin*, December 18, 1964.
3. W. D. Klein and L. M. Finnell, "Comparative Study of Coho Salmon Introductions in Parvin Lake and Granby Reservoir," *The Progressive Fish-Culturist* 31 no. 2; Delbert A. West, "Freshwater Silver Salmon, *Oncorhynchus kisutch* (Walbaum)," *Calif. Dept. Fish & Game* 51 (1965): 210–12, in Wayne H. Tody and Howard A. Tanner, Fish Division, Michigan Department of Conservation. *Coho Salmon for the Great Lakes*, Fish Management Report No. 1, February 1966. Arthur Whitney, personal communication, 1965, Montana Department of Fish and Game, Helena, in Wayne H. Tody and Howard A. Tanner, Fish Division, *Coho Salmon for the Great Lakes*, Fish Management Report No. 1, February 1966. P. Fuller, J. Larson, and A. Fusaro, *"Oncorhynchus kisutch," U.S. Geological Survey*, http://nas.er.usgs.gov/queries/FactSheet.aspx?speciesID=908 (rev. 6/26/2014).
4. R. E. Foerster and W. E. Ricker, "The Coho Salmon of Cultus Lake and Sweltzer Creek," *Journal of the Fisheries Research Board of Canada* 10 (1953): 293–319.
5. Orlay W. Johnson, Thomas A. Flagg, Desmond J. Maynard, George B. Milner, F. William Waknitz, "NOAA Technical Memorandum F/NWC-202 Status Review for Lower Columbia River Coho Salmon: June 1991," https://www.nwfsc.noaa.gov/assets/25/5969_06212004_124103_202.pdf.

THE HATCHERY SITUATION

1. Harry Westers and Thomas M. Stauffer, "The History of Fish Culture in Michigan," in *Michigan Department of Natural Resources, Fisheries Management Report No. 6, Michigan Fisheries Centennial Report—1873–1973, Fisheries Division—April 1974*.
2. Albert S. Hazzard and David S. Shetter, "Results from Experimental Plantings of Legal-Sized Trout," Institute of Fisheries Research, *Michigan Department of Conservation and University of Michigan, Report No. 480, June 1938* (American Fisheries Society).
3. "Watersmeet Trout Hatchery and Fish Farm," http://www.watersmeettrouthatchery.com/

about.html.

4. Michigan Department of Natural Resources, Fishing, "Platte River State Fish Hatchery," http://www.michigan.gov/dnr/0,4570,7-153-10364_52259_28277-22491--,00.html.

FINANCIAL CHALLENGES

1. *Anadromous Fish Conservation Act of 1965*, https://www.fws.gov/laws/lawsdigest/anadrom.html; "16 U.S. Code § 757—Utilization of State Services; Expenditure of Funds," Cornell University, Legal Information Institute, http://www.law.cornell.edu/uscode/text/16/757.
2. P. Fuller, J. Larson, and A. Fusaro, "*Oncorhynchus kisutch*," USGS Nonindigenous Aquatic Species Database, Gainesville, FL, https://nas.er.usgs.gov/queries/FactSheet.aspx?SpeciesID=908 (rev. 6/26/2014).

MAKING THE CASE

1. Howard A. Tanner, "Three Fine Fish," *Michigan Conservation* (March–April 1965).
2. Howard A. Tanner, "Great Lakes, Sport Fishing Frontier," *Michigan Conservation* (November–December 1965).
3. Wayne H. Tody and Howard A. Tanner, Michigan Fish Division, *Coho Salmon for the Great Lakes*, Fish Management Report No. 1, February 1966.

NEW PERSONNEL

1. David P. Borgeson, Reflections on the Early Salmon Years, personal correspondence, 2012.

THE SUMMER OF 1966

1. Wayne H. Tody, *A History of Michigan's Fisheries: A Fabulous Renewable Natural Resource along with Our Forests and Wildlife* (Traverse City, MI: Copy Central, 2003).
2. Wayne Tody to C. D. Harris and Dorias J. Curry, June 20, 1966.

THE JACK RUN

1. Michigan Department of Natural Resources, "Michigan's First Plantings of Coho Salmon," *Official News Bulletin*, March 24, 1966.

ALEWIVES REPRISE

1. Federal Water Pollution Control Administration, Great Lakes Region, "The Alewife Explosion: The 1967 Die-off in Lake Michigan," http://nepis.epa.gov/Exe/ZyPDF.cgi/2000UXW0.PDF?Dockey=2000UXW0.pdf.

2. Stevenson Swanson, "Smell of Money Might Make Alewives Tolerable," *Chicago Tribune*, May 5, 1992.

THE DREAM COME TRUE AND A NIGHTMARE

1. Tom McNally, "Cohos Show Up; So Do Anglers," *Chicago Tribune*, March 31, 1968.
2. David P. Borgeson, *Fish Management Report No. 3, Coho Salmon Status Report 1967–68* (Fisheries Division, Michigan Department of Natural Resources, February 1970).
3. Tom McNally, "Salmon Are in for Real at Manistee," *Chicago Tribune*, September 20, 1967.
4. William Gautsche, personal communication, 2012.
5. United Press International, "6 Fishermen Drowned in Lake Squall, Huge Waves Hit Small Craft," *Chicago Tribune*, September 24, 1967.
6. Tom McNally, "Sportsmen Find '67 Fouled by Weathermen," *Chicago Tribune*, December 24, 1967.

SWIMMING UPSTREAM

1. John MacGregor, personal communication, 2012.
2. Wayne H. Tody, personal communication.
3. Ibid.

SUSTAINING THE EXCITEMENT

1. Associated Press, "People in the News," *Gettysburg Times* (PA), September 4, 1968.
2. William Gautsche, personal communication, 2012.
3. Gordon Young, James L. Amos, and Martin Rogers, "The Great Lakes: Will Our Inland Seas Survive?" *National Geographic*, August 1973.
4. Journal of the House of Representatives, State of Michigan, Regular Session, 1968, vol. 1, House Concurrent Resolution No. 204.

STEWARDSHIP AND CONTAMINANTS

1. Thomas R. Dunlap, *DDT: Scientists, Citizens, and Public Policy* (Princeton, NJ: Princeton University Press), 1981.
2. Ibid.
3. State of Michigan, Department of Community Health, "Michigan Fish Consumption Advisory Program Guidance Document, Version 2," September 17, 2014, http://www.michigan.gov/documents/mdch/MDCH_MFCAP_Guidance_Document_417043_7.pdf.
4. *Environment: Science and Policy for Sustainable Development* 11, no. 3 (1969).

ECONOMICS

1. *Traverse City Record-Eagle*, October 5, 1967.
2. Department of Information Services, Michigan State University, "Coho's Impact on State Economy Aired at Session," October 15, 1968.
3. Ibid.
4. Ibid.
5. State of Michigan, Department of Treasury, State Equalized Valuations for 1966–1970 in Antrim, Benzie, Grand Traverse, Leelanau, Manistee, Mason and Oceana Counties: Calculation of Comparison between Lakeshore and Non-Lakeshore Townships.
6. "Outdoors: Coho Madness," *Time*, September 27, 1968.
7. Theodore Karamanski, *A Nationalized Lakeshore: The Creation and Administration of Sleeping Bear Dunes National Lakeshore* (National Park Service, U.S. Department of the Interior, 2000), https://www.nps.gov/parkhistory/online_books/slbe/.
8. Michigan Department of Natural Resources, *Michigan Harbors Guide*, http://www.michigandnr.com/Publications/PDFS/RecreationCamping/boating/harbor_guide_small.pdf.
9. "Evening News Outdoors," *Evening News*, Sault Saint Marie, Michigan, October 29, 1969.
10. "Boating 1998," chart prepared by the National Marine Manufacturers Association, Chicago, Illinois, in Ryck Lydeker and Margaret Podlich, Boat Owners Association of the United States, "A Profile of Recreational Boating in the United States," in *Trends and Future Challenges for U.S. National Ocean and Coastal Policy*, http://oceanservice.noaa.gov/websites/retiredsites/natdia_pdf/14boatus.pdf.
11. Michigan Sea Grant, *Economic Impacts of Charter Fishing in Michigan, 2013*, http://www.miseagrant.umich.edu/explore/fisheries/economic-impacts-of-charter-fishing-in-michigan/.
12. *Traverse City Record-Eagle*, October 5, 1967.
13. Michigan Sea Grant, *Economic Impacts of Charter Fishing in Michigan, 2013*.
14. "Great Lakes Restoration at Work in Michigan," http://www.allianceofrougecommunities.com/PDFs/rrac/State%20Factsheet_Michigan_Feb%2025_2011_final.pdf.

INDEX

A

advisories, fish consumption guidance, 231

alewives, 107, 109, 110, 231; and beaches, 123, 233; commercial fishing of, 151, 203, 204, 220; dying off of, 122, 196, 202, 203, 206; introduction of, 74, 148, 175, 202; life cycle of, 174; population of, xviii, 120, 121, 122, 201, 241; and salmon, 120, 123, 130, 138, 143, 154, 163, 241; and sea lamprey, 179; Tanner on, 207

American Fisheries Society (AFS), 130, 177, 183, 225

ammocetes, 117–18

Anadromous Fish Act, 173, 188

Applegate, Vernon, 29, 117–18

Arkansas River, 41, 47, 48

Army, U.S., 10, 13, 15, 17, 21, 25, 39

Asian carp, 241. *See also* carp

Atlantic salmon, 66, 69–70, 115, 148, 243

Au Sable River, 155–56

B

Ball, Robert, 25–31, 33–35; calling about salmon eggs, 137; leader in developing limnology and fisheries science, 73; recruiting of Tanner as fish chief, 86; recruiting of Tanner to return to MSU, 190; supporting of Tanner and Helen, 89

Bay de Noc, 96

Bay Mills Indian Community, 101, 102, 104

beaches, 15; alewife nuisance on, 122, 123, 196; grappling with carcasses on, 201–2, 204, 215, 233

Bear Creek, 168, 188–89, 194, 215, 217

Beaver Island, 116

Beckman, William, 38–40, 45, 153

Bellaire, 4; brook trout fishing in, 6, 8, 11; Tanner growing up in, 14, 16, 33, 128

benthic zone, 109, 241. *See also* zones

Beulah, 69–70

Big Huron River, 175

Big Manistee River, 168, 217

biologists, 40, 50, 57, 149, 207, 218; Bryan Burrows, 159; Colorado, 44, 48; Dick Klein, 51; fish divisions and, 82; fishery, 177; introducing species, 69, 74, 126, 143, 203; opinions on sea lamprey impact, 113; Oregon and Washington, 140, 187, 194, 205; and reservoirs, 42, 52; and resource management, 41, 108, 123; smelt and, 69–70; staff, 163, 167; and tribes, 104; Vernon Applegate, 117–18; Walt Crowe, 28; Wayne Tody, 162, 197, 198

biology, 35, 37, 44–45, 72, 74, 84, 87, 153; food and food web, 73, 109; in freshwater environments, 37, 240; and the Great Lakes, 142, 175; growth of, 36, 87; of lake trout, 64, 107, 108; of salmon, 138, 160; of sea lamprey, 113; of striped bass, 178

Bonneville Power Authority, 141

Borgeson, Dave, 183

British Columbia, 52, 83, 95, 127, 164, 205

brook trout, 3–5, 17, 32, 42, 69, 79, 110, 149

brown trout, 32, 42, 69, 73, 79, 148, 155, 167

C

California, 24, 52, 73, 125, 138, 142, 164, 236

Canada, 54, 57, 60; commercial fishing in, 151; Great Lakes Fishery Commission, 98–99, 108; introductions of fish, 147–48; Isle Royale to United States, 65; Ministry of Fisheries and Oceans, 152; percentage of lakes, 61, 62; pink salmon experiment, 125, 188; public's waters, 92; research laboratories, 71, 79; sea lamprey control, 117, 119

canals, 66, 71, 115, 121, 148

Carbine, Fenton, 153–54, 204, 231

carp, 72, 147; Asian, 241; European, 42, 49, 69, 72, 147

Carp River, 117

Carson, Rachel, 230

Charlevoix, 29, 138, 183

Chicago, 201, 223–24, 233

chinook salmon (also king): critical factor, xviii; description, nomenclature, and advantages of, 130–31; eggs from Washington, 193; introduction of other new species, 243; in Michigan hatchery, 198; natural reproduction of, 241; a Pacific salmon, 124

chubs, 63, 94, 95

Coast Guard, U.S., 214, 224

Cochrane Lake (also North Twin), 27, 30, 38

coho salmon (also silver), 177–80, 218, 231–34, 237, 241; Carbine opposition of, 153–54; chinook advantages over, 131; Colorado and, 45, 52; critical factors and, xviii; eggs, 51, 107, 137–38, 198; fishing of, 205–6, 209–11, 215–16; historical experiences with, 124, 125; introduction of, 187–89, 220, 225; management of, 142–43, 182, 243; news coverage of, 194–96, 222–23; nomenclature of, 130; Oregon and Washington and, 43, 146–47, 161–66; pink salmon history and, 126; questions and caution introducing, 53, 149–50; sea lamprey control and, 118;

spawning of, 133

Colorado: court battles with, attorneys, 102; critical factor, xviii, 37; kokanee and coho from, 142, 164, 166, 127–28, 137–38; state description, 40–54; Tanner's experiences in, 79–80, 85, 89, 140, 153, 191–92; trout in, 155

Colorado Agricultural and Mechanical College (now Colorado State University), 26, 38–40; Colorado Cooperative Fishery Research Unit, 38, 40, 45, 47, 50, 53, 153

Colorado Division of Wildlife, 41

Colorado Fisheries Research, 47, 51, 53

Colorado Game, Fish and Parks, 47, 49

Colorado River, 41, 48, 51

Columbia River: condition of, 140–41, 143, 149, 164, 192, 194; decline of coho in, 52, 137

Congress, United States, 62, 67, 68, 103, 174

conservation, principles of, 80, 167

courts: U.S. Appeals, 102, 103; Michigan Supreme, 96

critical factors: accidental introduction of alewife, 122; changes in fishery key value, 93; changes in Michigan government, 80; coho salmon in Colorado, 53; gaining control over sea lamprey, 119; Great Lakes habitat, 63; Michigan Department of Conservation leadership, 146; Ohio's refusal to give Great Lakes Fishery Commission management authority, 98; Oregon's Moist Pellet food, 140; pink salmon success, 125; public demand for recreation, 79; public support for the plan, 204; relationship with West Coast fishery biologists, 140; science of fisheries

management, 72

Crowe, Walt, 28–30, 117

Crystal Lake, 69, 70, 128

Cultus Lake, 164

cutthroat trout, 42, 48, 149

cycles, life. *See* life cycles

D

dams, hydroelectric, 140, 163–64, 193

DDT (Dichlorodiphenyltrichloroethane), 62, 230–31

Detroit River, 80

dissolved oxygen, 30, 32, 34

Donelson, Loren, 193, 198

E

East, Ben, 80

East Coast, 65, 67, 68, 122

Eisenhower, Dwight, 47

Empire, 207, 211, 214–15, 233

Engstrom, Arnell, 188

epilimnion, 35, 36. *See also* zones

Erie Canal, 67, 71, 115, 121

F

fertilization, 30–35, 143

fertilizers, 26–28, 30–31, 34–35, 37; inorganic, 27–28; potash, 28

fisheries, 24, 30; alewives, 122; biologists and, 28, 40, 82; biology of, 72; coho and trout, 177; in Colorado, 87, 121; commercial, 62; cooperative, 100, 150, 239; GLFC authority and, 98, 99, 100, 152; Great Lakes, 64, 107, 152, 241, 242; hatchery, 241; inexperienced, 182, 192; inland, 142, 151, 167, 173, 182, 183; invasive species in, 243; lake trout, 107, 108, 109, 110; management, 72, 73, 74, 80, 81, 84, 112; in Michigan, xviii, 68, 87, 88, 91, 93, 96, 102, 103, 239, 242; Native American, 103, 104; research, 38, 40, 53

Fort Collins, 39, 45, 51, 86–87, 89

Frankfort, 69, 206–7, 214–15, 224, 233

fry, 70, 82, 109–10

G

Gautsche, Bill, 212, 222

gear, 141, 146, 213, 218; compass, 213; downrigger, 212, 235; finders, 213; fishing, 68, 94–95, 102, 109, 220; hooks, 33, 131, 210–11, 223; horsepower, 208, 210, 236; Luhr Jensen, 221; lures, 211, 212, 225, 235; outboard, 234, 236; radios, 12, 21, 213, 235–36; reels, 208, 235; rods, 235

Georgetown Reservoir, 52, 142

gill nets. *See* nets

glaciers, 34, 42, 57, 107

Golden Gate, 14, 22

Governor's Special Conservation Study (Blue Ribbon), xviii, 80, 155

Granby Reservoir, 51–52, 87, 138

Grand Rapids, 14, 61, 156, 207

grayling, 49, 67

Great Depression, 11, 44, 91

Great Lakes Fishery Commission (GLFC): commercial fishing orientation, 151; composition of, 152; contrast with Trout Unlimited support, 154; critical factor, xviii; discovery of TFM, 118; management authority over Great Lakes, 98; management of alewife, 122, 204; McFadden recommendations about, 84,

92; Ohio refuses management authority, 99–100; organization of, 150; tribes on lake committees of, 104

Great Lakes salmon, xix, 53, 93, 124, 155, 205, 208, 225

Great Lakes Sport Fishing Council, 225

Green River, 41

Griffith, George A., 81

Gunnison River, 51

H

habitats: chinook, 130; critical factor, xviii, 33, 36, 44, 57, 115; destruction of, 66–67, 241; Great Lakes freshwater, 63, 73; hatchery, 192; introduction, 124; lake and stream restoration, 83–84; Lake Superior, 110; lake trout, 8, 49; 63, 64, 69, 70, 91, 107–21, 148, 152, 159, 179, 196

Hannah, John, 80, 87, 138, 164, 191, 222

harbors, 12, 15–16, 18, 22, 67, 116, 234

Harris, Chuck, 88, 128, 145–46, 154, 187

hatcheries, 26–27; Blue Ribbon committee critique of, 80, 82, 83, 84, 109, 110–11; California salmon, 52, 69, 72–73; Colorado, 52; Colorado experience with, 42, 44–45, 47, 49; eggs from Colorado in, 138; Marquette, 111; Oden, 177; Oregon, 141–43, 151, 158; Platte River, 169, 188, 194; pink salmon in, 125, 126, 128, 131; rainbow trout from, 75; review of Michigan, 161–69, 176–77, 182, 187–88, 192–93, 198, 217–18, 241; Watersmeet, 168

Higgins Lake, 30, 128–29, 150, 178

Hollandia, 15–16

Honor, 168, 188, 207, 237

hooks, 33, 131, 210–11, 223

horsepower, 208, 210, 236. *See also* gear

Horsetooth Reservoir, 41, 51

House Appropriations Committee (Michigan), 188

Hudson Bay, pink salmon in, 125–26

Hudson River, 66–67, 71, 115

hydroelectric dams, 140, 163, 193

hypolimnion, 36–37. *See also* zones

I

Illinois, 13, 39, 61, 79, 214–15, 220–21

Indian River, 26–28

Indiana Department of Natural Resources, 206

Institute for Fisheries Research (IFR), 27–28, 31, 38, 82–83, 88, 153

Intermediate River, 6–8

Isle Royale, 65, 68

J

jack run, 194–98, 205–6, 217

Jefferies, Ernie, 140, 196

Jensen, Luhr, 221

Johnson, Carl, 188

Jordan River, 5, 9, 69, 70, 110–11, 138, 151, 158

Jordan River Valley, 4–5, 110, 158

K

Kalamazoo, 9, 13, 22, 212, 235

Keller, Myrl, 183

king salmon, 130, 237. *See also* chinook salmon

Klein, Dick (W. D.), 51–53, 138

kokanee salmon, 51–52, 88, 124, 126–29, 137, 142, 153, 164, 166, 176, 178–79, 187

L

Lake and Stream Improvement Section (LSIS), 82, 83

Lake Champlain, 115

Lake Erie, 67, 100, 116, 178

Lake Huron, 10, 57, 60, 62, 65, 67, 107, 108, 113, 116, 117, 119, 121, 126, 167, 175, 178, 234

Lake Michigan, 7, 60, 61, 69, 87, 96, 98, 109, 116, 119, 122, 129, 138, 159, 167, 175, 180, 196, 201–4, 206–7, 213–14, 220, 231, 233, 237

Lake Nipigon, 125

Lake Ontario, xvii, 62, 66, 71, 108, 115, 121, 223

Lake St. Clair, 60, 67, 183, 196

Lake Superior, xvii, xviii, 60, 65, 108, 111, 117–19, 121, 125, 126, 142, 167, 175, 187

lake trout, 26, 31, 33, 35, 49, 63–64, 69–70, 91, 107–21, 148, 151–52, 159, 179, 196

lamprey, sea. *See* sea lamprey

Lansing, 30, 67, 81, 89, 159, 183

LeBlanc, Albert (Big Abe), 102

legislature, 62, 68, 84, 152, 161, 174, 179–80, 187–88, 234, 243

Leyte (island), 15, 18–19

license: commercial fishing, 93, 95, 97, 101–2; fishing, 63

life cycles, 34, 52, 120, 122, 124, 138, 149, 164, 174

Little Manistee River, 156, 217–18

Lower Peninsula, 69, 111, 206, 208, 233

lures. *See* gear: lures

M

MacGregor, John, 218–19, 221–22

Mack, Joe, 175

Mackinaw City, 117, 122

MacMullan, Ralph A., 30, 81, 83–84, 89, 128, 137, 145–46, 174–75, 180, 204, 206, 231

Maine, 69, 70

Mancelona, 3–4, 6, 11, 70

Manistee, 104, 159, 204, 206–7, 210–11, 213, 221–22, 232–33

Manistee Lake, 159, 211, 213

Manistee River, 159, 175, 195–97, 206, 215, 218

Marquette Fish Hatchery, 111

Mason Tract, 155–56

McFadden, James, 83, 89, 141, 145

Menominee River, 138

mercury, 62

Michigan Conservation, 178

Michigan Conservation Commission, 44, 70, 80–81, 84, 90–94, 96, 101, 103–4, 142, 145, 153–55, 158, 167, 174, 177, 188, 219, 231, 244

Michigan Conservation Department (also Department of Conservation), xvii, xviii, 5, 26–27, 31–32, 35, 38, 75, 80–82, 88, 92, 96, 111, 129, 131, 154–56, 169, 175, 180, 187–89, 197–98, 202, 209–11, 213, 216–17, 234–35, 240

Michigan Department of Natural Resources, 68, 81, 129, 130, 169, 180, 189, 197–98, 202, 209, 240

Michigan Fish Commission, 68

Michigan Fish Division, xvii, 70, 75, 80–84, 88, 90, 94–96, 100, 102, 107, 111, 123, 127, 137, 142, 146, 152, 154, 156–59, 161–63, 165–66, 174, 179, 182–83, 189–91, 193, 197–98, 208, 216–17, 219–20, 231

Michigan Natural Resources Commission, 235

Michigan Sea Grant, 236

Michigan State College, 9–10, 14, 24, 59, 80, 86–87, 137, 189, 190, 222; Center for Economic Analysis, 236; College of Agriculture, 190, 222, 230; Department of Fisheries and Wildlife, 27, 85, 89, 190; Director of Natural Resources, 27, 190, 230; Michigan State University (MSU), 27, 190, 222, 230, 236

Michigan Steelhead and Salmon Fishermen's Association (MSSFA), 103

Michigan Territory, history of, 65–68

Michigan United Conservation Clubs (MUCC), 103, 225

Millenbach, Cliff, 140, 161, 196

Mindoro, 19–20

Ministry of Fisheries and Oceans Canada, 152

Minnesota, 61, 126, 150

Mio Reservoir, 155

Missouri, 14, 39

Montana, 52, 125, 138, 142, 155, 164

Morrill Hall, 25, 117

Mullendore, Bill, 128, 161, 187

mussels, 109, 241

N

National Geographic, 222

national parks, 233

Native American Rights Fund (NARF), 102

Native Americans, 64–65, 103, 125

natural reproduction, 84, 162–63, 167–68, 241

Neff, Mort, 195

nets, 15, 21, 46, 66–67, 70, 91, 94–96, 101–2, 104, 108–10, 113, 116, 151, 196, 213; gill, 44, 66, 94–95, 101–2, 116; nylon, 91, 113, 151, 212; trawl, 122, 179, 204

New York, 60–61, 115, 223

Newton, Lyle, 168

Niagara Falls, 11

nitrogen, 28

non-indigenous species, 61, 124, 147, 150

North America, 34, 57, 64–65, 72, 74, 102, 115, 124, 127, 147–48, 179

North Twin Lake (also Cochrane), 27, 30, 38

Northwest Territory, 65–66

NPK, 28. *See also* fertilizers

O

Oden, 177

Ohio, xviii, 26, 61, 67–68, 99–100, 183

Ontario, xxii, 61, 62, 66, 68, 71, 84, 99–100, 115, 117, 121, 125, 223

Oregon, 137, 138, 140, 141, 143, 144, 146, 147, 149, 150, 161, 163, 165, 166, 168, 177, 179, 180, 187, 192, 194, 196, 198, 205

Oregon Moist Pellet, 140, 141, 147, 153, 164–66, 173, 180, 188

overfishing, 92, 121

oxygen, 30, 34–37, 130; dissolved, 30, 32, 34

P

Pacific Islands, 29

Pacific Ocean, 15, 124, 182

Pacific salmon, xvii, 57, 63, 69, 70, 100, 111, 119, 120, 124, 125, 140, 142, 146, 148, 149, 154, 165, 179, 202, 235, 239, 241

panfish, 63

parks, 42, 44, 47, 49, 51, 53, 79, 86, 190, 233

Parsons, John, 148

PCBs (polychlorinated biphenyls), 62, 230

pelagic, 109, 130, 163, 241. *See also* zones

Pentwater, 206, 233

Pere Marquette River, 138

phosphorus, 28. *See also* fertilizers

phytoplankton, 37, 51. *See also* plankton

Pigeon River, 28, 35; State Forest, 31

pink salmon, xviii, 124–26, 142, 148, 188

plankton, 28, 34, 36–37, 51, 129, 133, 241; phytoplankton, 37, 51; zooplankton, 36, 37, 51

Platte Bay, fishing trips with friends at, 207–9, 211, 214–15

Platte River, 41, 168–69, 175, 188, 194–95, 207–8, 215, 217–18, 222, 225, 237

policies, 62, 80, 82, 92, 145, 150–51, 241–43

polychlorinated biphenyls (PCBs). *See* PCBs (polychlorinated biphenyls)

potash, 28. *See also* fertilizers

Q

quagga, 241

Quebec, 62

R

radios, 12, 21, 213, 235–36. *See also* gear

rainbow smelt, 63, 70–71, 74. *See also* smelt

rainbow trout, 5–7, 9, 42, 45, 69, 73, 132, 149, 167

recreation, 39, 83, 155, 175, 190, 202

reefs, 17, 109–10

reels, 208, 235. *See also* gear

reservoirs, 40–46, 48–52, 74, 87, 142. *See also* Georgetown Reservoir; Granby Reservoir; Horsetooth Reservoir; Mio Reservoir

Ricker, William, 164

Rio Grande, 42, 48

Robertson, Russ, 111

Romney, George, 80–81, 128–29, 153–54, 174, 188, 222

rotenone, 32, 43

R/V *Tanner*, 225

S

Sacramento River, 73

salmon (Salmonidae), 93, 100, 107, 112, 114, 118, 121, 123, 124–25, 130, 132, 137–38, 140–42, 149, 151, 152, 153. *See also* Atlantic salmon; chinook salmon (also king); coho salmon (also silver); Great Lakes salmon; king salmon; kokanee salmon; Pacific salmon; pink salmon; silver salmon; sockeye salmon

San Francisco, 14, 22, 39

Saturday Evening Post, xvii, 222

sea lamprey: accidental introduction of, 71, 74, 147–48; and alewife, 121, 150–51; Anadromous Fish Act and, 174; assessment of spawning habitat, 29, 57; control of, 179; detailed description of, 113–19; GLFC management of, 84; impact on commercial fishery, 91; and lake trout, 109–11; number of species of, 107

Senate Appropriations Committee (Michigan), 175

shorelines, 17, 31, 50, 60–61, 74, 202, 204, 215, 232, 234, 237

Silent Spring, 230

silver salmon, 18, 130, 147, 211, 212, 220. *See also* coho salmon (also silver)

smelt, 9, 63, 69, 70, 71, 109, 110, 133, 148, 174; rainbow, 63, 70–71, 74

smolts, 129, 130, 141, 149, 162, 163, 164, 167, 175, 180, 189, 194, 217, 219, 220, 225

sockeye salmon, 127

species, 5, 43, 44, 52, 63, 68, 73, 91, 94, 100,

107, 149; anadromous, 173; commercial, 94, 95, 107, 109, 110, 111, 112, 151; introduced, 8, 9, 32, 42, 48, 49, 50, 51, 52, 54, 61, 63, 69, 70, 72, 74, 75, 113, 114, 115, 116, 122, 124, 125, 126, 130, 140, 142, 143, 147, 148, 150, 152, 164, 178, 203; invasive, 241; native, 3, 32, 64, 67, 68, 69, 73, 107, 109, 114, 115, 116, 151, 159, 167; sport, 126, 144, 147, 152

sport fishing, 109, 179, 225

Sport Fishing Institute, 225

Steelhead, S/V, 133, 183

steelhead trout, 63, 73, 103, 131–33, 149, 194

stewardship, 229, 231, 243

St. Joe, 138

St. Lawrence River, 61, 11

St. Lawrence Seaway, 242

stock, 21, 32, 43, 46, 48, 49, 51, 52, 69, 82, 84, 92, 93, 95, 97, 99, 107, 109, 110–11, 119, 124, 126–27, 129, 138, 142, 148, 162–63, 166–67, 175–77, 189, 192, 219

storms, 214, 224, 234, 242

suckers, 8, 32, 52

S/V *Steelhead*, 133, 183

T

Tack, Peter, 9–10, 25–26, 85–86, 88, 190

tackle, 7, 131–32, 174, 196, 210, 221, 223, 235; Helin Tackle Company, 235. *See also* gear

Tanner, Helen, 12–14, 22, 24–25, 28–29, 191, 222, 225

Tanner, R/V, 225

taxes, 96, 174, 233–34, 241, 243

TFM (trifloromethyl-4-nitrophenol), 118

thermocline, 36, 37. *See also* zones

Tody, Wayne, xviii, 83, 89, 142, 154, 158, 162,

173, 177–78, 179, 182–83, 192, 194–97, 204, 206

Torch Lake, 13, 116, 128–29

Toronto, 11, 61

tourists, 122, 190, 201–2, 208, 232–33, 236, 237

Traverse City Record-Eagle, 195, 223

treaties, 61, 65, 101–3

tribes, 101–4

tributaries, 5, 29, 61, 66, 70, 73, 115–17, 125–26, 128–30, 138, 163, 168, 175, 189, 217

trout. *See* brook trout; brown trout; cutthroat trout; lake trout; rainbow trout

Trout Unlimited, 154–59, 225

U

United States Bureau of Commercial Fisheries, 71, 73, 82, 98–100, 107, 110, 117–18, 122–23, 152–54

United States Bureau of Indian Affairs, 102

United States Bureau of Reclamation, 41, 47, 49

United States Congress, 67

United States Court of Appeals, 103

United States Fish and Wildlife Service, 40, 99, 153, 204

United States Fish Commission, 73

University of Michigan, 24, 26, 71, 82, 153

University of Washington, 193, 198

Upper Manistee River, 156, 218

Upper Peninsula, 17, 65, 68–69, 96, 122, 138, 168, 175, 219

U.S. Army, 10, 13, 15, 17, 21, 25, 39

V

VerDuin, Claude, 151

W

Wallace, George, 230

Washington, DC, 153

Washington (state), 137–38, 140–41, 143–44, 146, 149–50, 161, 163, 177, 179, 192, 194, 198, 205

watershed, 8, 34, 69, 111, 115, 156

Watersmeet hatchery, 168. *See also* hatcheries

Waterways Division, 234

Welland Canal, 71, 115, 121

West Coast, 140, 142, 149, 163, 165, 173, 179, 183, 187, 192, 195, 197, 208, 221

Western Michigan College, 9, 12

White River, 41, 50

whitefish, 9, 48, 63

Wilson, Keith, 234

winterkill, 30, 34

Wisconsin, 61, 73, 150, 168, 204, 214, 220–21, 231

Woodward, Ernie, 47

Z

zones: benthic, 109, 241; epilimnion, 35–36; hypolimnion, 36–37; pelagic, 109, 130, 163, 241; thermocline, 36–37

zooplankton, 36–37, 51. *See also* plankton